# Doing Data Science

*Cathy O'Neil and Rachel Schutt*

Beijing · Cambridge · Farnham · Köln · Sebastopol · Tokyo

**Doing Data Science**

by Cathy O'Neil and Rachel Schutt

Published by O'Reilly Media, Inc., 1005 Gravenstein Highway North, Sebastopol, CA 95472.

O'Reilly books may be purchased for educational, business, or sales promotional use. Online editions are also available for most titles (*http://safaribooksonline.com*). For more information, contact our corporate/institutional sales department: 800-998-9938 or *corporate@oreilly.com*.

**Editors:** Mike Loukides and Courtney Nash
**Production Editor:** Kristen Brown
**Copyeditor:** Kim Cofer
**Proofreader:** Amanda Kersey

**Indexer:** WordCo Indexing Services
**Cover Designer:** Karen Montgomery
**Interior Designer:** David Futato
**Illustrator:** Rebecca Demarest

October 2013:     First Edition

**Revision History for the First Edition:**

2013-10-08:   First release

2013-12-13:   Second release

2014-10-10:   Third release

See *http://oreilly.com/catalog/errata.csp?isbn=9781449358655* for release details.

ISBN: 978-1-449-35865-5

[LSI]

*In loving memory of Kelly Feeney.*

# Table of Contents

# Preface

*Rachel Schutt*

Data science is an emerging field in industry, and as yet, it is not well-defined as an academic subject. This book represents an ongoing investigation into the central question: "What is data science?" It's based on a class called "Introduction to Data Science," which I designed and taught at Columbia University for the first time in the Fall of 2012.

In order to understand this book and its origins, it might help you to understand a little bit about me and what my motivations were for creating the class.

## Motivation

In short, I created a course that I wish had existed when I was in college, but that was the 1990s, and we weren't in the midst of a data explosion, so the class couldn't have existed back then. I was a math major as an undergraduate, and the track I was on was theoretical and proof-oriented. While I am glad I took this path, and feel it trained me for rigorous problem-solving, I would have also liked to have been exposed then to ways those skills could be put to use to solve real-world problems.

I took a wandering path between college and a PhD program in statistics, struggling to find my field and place—a place where I could put my love of finding patterns and solving puzzles to good use. I bring this up because many students feel they need to know what they are "going to do with their lives" now, and when I was a student, I couldn't plan to work in data science as it wasn't even yet a field. My advice to students (and anyone else who cares to listen): you don't need to figure it all out now. It's OK to take a wandering path. Who knows what you might find? After I got my PhD, I worked at Google for a few years around the same time that "data science" and "data scientist" were becoming terms in Silicon Valley.

The world is opening up with possibilities for people who are quantitatively minded and interested in putting their brains to work to solve the world's problems. I consider it my goal to help these students to become critical thinkers, creative solvers of problems (even those that have not yet been identified), and curious question askers. While I myself may never build a mathematical model that is a piece of the cure for cancer, or identifies the underlying mystery of autism, or that prevents terrorist attacks, I like to think that I'm doing my part by teaching students who might one day do these things. And by writing this book, I'm expanding my reach to an even wider audience of data scientists who I hope will be inspired by this book, or learn tools in it, to make the world better and not worse.

Building models and working with data is not value-neutral. You choose the problems you will work on, you make assumptions in those models, you choose metrics, and you design the algorithms.

The solutions to all the world's problems may not lie in data and technology—and in fact, the mark of a good data scientist is someone who can identify problems that *can* be solved with data and is well-versed in the tools of modeling and code. But I do believe that interdisciplinary teams of people that include a data-savvy, quantitatively minded, coding-literate problem-solver (let's call that person a "data scientist") could go a long way.

## Origins of the Class

I proposed the class in March 2012. At the time, there were three primary reasons. The first will take the longest to explain.

**Reason 1**: I wanted to give students an education in what it's like to be a data scientist in industry and give them some of the skills data scientists have.

I was working on the Google+ data science team with an interdisciplinary team of PhDs. There was me (a statistician), a social scientist, an engineer, a physicist, and a computer scientist. We were part of a larger team that included talented data engineers who built the data pipelines, infrastructure, and dashboards, as well as built the experimental infrastructure (A/B testing). Our team had a flat structure.

Together our skills were powerful, and we were able to do amazing things with massive datasets, including predictive modeling, prototyping algorithms, and unearthing patterns in the data that had huge impact on the product.

We provided leadership with insights for making data-driven decisions, while also developing new methodologies and novel ways to understand causality. Our ability to do this was dependent on top-notch engineering and infrastructure. We each brought a solid mix of skills to the team, which together included coding, software engineering, statistics, mathematics, machine learning, communication, visualization, exploratory data analysis (EDA), data sense, and intuition, as well as expertise in social networks and the social space.

To be clear, no one of us excelled at all those things, but together we did; we recognized the value of all those skills, and that's why we thrived. What we had in common was integrity and a genuine interest in solving interesting problems, always with a healthy blend of skepticism as well as a sense of excitement over scientific discovery. We cared about what we were doing and loved unearthing patterns in the data.

I live in New York and wanted to bring my experience at Google back to students at Columbia University because I believe this is stuff they need to know, and because I enjoy teaching. I wanted to teach them what I had learned on the job. And I recognized that there was an emerging data scientist community in the New York tech scene, and I wanted students to hear from them as well.

One aspect of the class was that we had guest lectures by data scientists currently working in industry and academia, each of whom had a different mix of skills. We heard a diversity of perspectives, which contributed to a holistic understanding of data science.

**Reason 2**: Data science has the potential to be a deep and profound research discipline impacting all aspects of our lives. Columbia University and Mayor Bloomberg announced the Institute for Data Sciences and Engineering in July 2012. This course created an opportunity to develop the theory of data science and to formalize it as a legitimate science.

**Reason 3**: I kept hearing from data scientists in industry that you can't teach data science in a classroom or university setting, and I took that on as a challenge. I thought of my classroom as an incubator of data science teams. The students I had were very impressive and are turning into top-notch data scientists. They've contributed a chapter to this book, in fact.

## Origins of the Book

The class would not have become a book if I hadn't met Cathy O'Neil, a mathematician-turned-data scientist and prominent and out-spoken blogger on *mathbabe.org*, where her "About" section states that she hopes to someday have a better answer to the question, "What can a nonacademic mathematician do that makes the world a better place?" Cathy and I met around the time I proposed the course and she was working as a data scientist at a startup. She was encouraging and supportive of my efforts to create the class, and offered to come and blog it. Given that I'm a fairly private person, I initially did not feel comfortable with this idea. But Cathy convinced me by pointing out that this was an opportunity to put ideas about data science into the public realm as a voice running counter to the marketing and hype that is going on around data science.

Cathy attended every class and sat in the front row asking questions, and was also a guest lecturer (see Chapter 6). As well as documenting the class on her blog, she made valuable intellectual contributions to the course content, including reminding us of the ethical components of modeling. She encouraged me to blog as well, and so in parallel to her documenting the class, I maintained a blog (*http://columbiadatascience.com/blog/*) to communicate with my students directly, as well as capture the experience of teaching data science in the hopes it would be useful to other professors. All Cathy's blog entries for the course, and some of mine, became the raw material for this book. We've added additional material and revised and edited and made it much more robust than the blogs, so now it's a full-fledged book.

## What to Expect from This Book

In this book, we want to both describe and prescribe. We want to *describe* the current state of data science by observing a set of top-notch thinkers describe their jobs and what it's like to "do data science." We also want to *prescribe* what data science could be as an academic discipline.

Don't expect a machine learning textbook. Instead, expect full immersion into the multifaceted aspects of data science from multiple points of view. This is a survey of the existing landscape of data science—an attempt to map this emerging field—and as a result, there is more breadth than depth in some cases.

This book is written with the hope that it will find itself into the hands of someone—you?—who will make even more of it than what it is, and go on to solve important problems.

After the class was over, I heard it characterized as a holistic, humanist approach to data science—we did not just focus on the tools, math, models, algorithms, and code, but on the human side as well. I like this definition of humanist: "a person having a strong interest in or concern for human welfare, values, and dignity." Being humanist in the context of data science means recognizing the role your own humanity plays in building models and algorithms, thinking about qualities you have as a human that a computer does not have (which includes the ability to make ethical decisions), and thinking about the humans whose lives you are impacting when you unleash a model onto the world.

## How This Book Is Organized

This book is organized in the same order as the class. We'll begin with some introductory material on the central question, "What is data science?" and introduce the data science process as an organizing principle. In Chapters 2 and 3, we'll begin with an overview of statistical modeling and machine learning algorithms as a foundation for the rest of the book. Then in Chapters 4–6 and 8 we'll get into specific examples of models and algorithms in context. In Chapter 7 we'll hear about how to extract meaning from data and create features to incorporate into the models. Chapters 9 and 10 involve two of the areas not traditionally taught (but this is changing) in academia: data visualization and social networks. We'll switch gears from prediction to causality in Chapters 11 and 12. Chapters 13 and 14 will be about data preparation and engineering. Chapter 15 lets us hear from the students who took the class about what it was like to learn data science, and then we will end by telling you in Chapter 16 about what we hope for the future of data science.

# How to Read This Book

Generally speaking, this book will make more sense if you read it straight through in a linear fashion because many of the concepts build on one another. It's also possible that you will need to read this book with supplemental material if you have holes in your probability and statistics background, or you've never coded before. We've tried to give suggestions throughout the book for additional reading. We hope that when you don't understand something in the book, perhaps because of gaps in your background, or inadequate explanation on our part, that you will take this moment of confusion as an opportunity to investigate the concepts further.

# How Code Is Used in This Book

This isn't a how-to manual, so code is used to provide examples, but in many cases, it might require you to implement it yourself and play around with it to truly understand it.

# Who This Book Is For

Because of the media coverage around data science and the characterization of data scientists as "rock stars," you may feel like it's impossible for you to enter into this realm. If you're the type of person who loves to solve puzzles and find patterns, whether or not you consider yourself a quant, then data science is for you.

This book is meant for people coming from a wide variety of backgrounds. We hope and expect that different people will get different things out of it depending on their strengths and weaknesses.

- Experienced data scientists will perhaps come to see and understand themselves and what they do in a new light.
- Statisticians may gain an appreciation of the relationship between data science and statistics. Or they may continue to maintain the attitude, "that's just statistics," in which case we'd like to see that argument clearly articulated.
- Quants, math, physics, or other science PhDs who are thinking about transitioning to data science or building up their data science skills will gain perspective on what that would require or mean.

- Students and those new to data science will be getting thrown into the deep end, so if you don't understand everything all the time, don't worry; that's part of the process.

- Those who have never coded in R or Python before will want to have a manual for learning R or Python. We recommend *The Art of R Programming* by Norman Matloff (No Starch Press). Students who took the course also benefitted from the expert instruction of lab instructor, Jared Lander, whose book *R for Everyone: Advanced Analytics and Graphics* (Addison-Wesley) is scheduled to come out in November 2013. It's also possible to do all the exercises using packages in Python.

- For those who have never coded at all before, the same advice holds. You might also want to consider picking up *Learning Python* by Mark Lutz and David Ascher (O'Reilly) or Wes McKinney's *Python for Data Analysis* (also O'Reilly) as well.

# Prerequisites

We assume prerequisites of linear algebra, some probability and statistics, and some experience coding in any language. Even so, we will try to make the book as self-contained as possible, keeping in mind that it's up to you to do supplemental reading if you're missing some of that background. We'll try to point out places throughout the book where supplemental reading might help you gain a deeper understanding.

# Supplemental Reading

This book is an overview of the landscape of a new emerging field with roots in many other disciplines: statistical inference, algorithms, statistical modeling, machine learning, experimental design, optimization, probability, artificial intelligence, data visualization, and exploratory data analysis. The challenge in writing this book has been that each of these disciplines corresponds to several academic courses or books in their own right. There may be times when gaps in the reader's prior knowledge require supplemental reading.

*Math*
- *Linear Algebra and Its Applications* by Gilbert Strang (Cengage Learning)

- *Convex Optimization* by Stephen Boyd and Lieven Vanden-berghe (Cambridge University Press)
- *A First Course in Probability* (Pearson) and *Introduction to Probability Models* (Academic Press) by Sheldon Ross

*Coding*
- *R in a Nutshell* by Joseph Adler (O'Reilly)
- *Learning Python* by Mark Lutz and David Ascher (O'Reilly)
- *R for Everyone: Advanced Analytics and Graphics* by Jared Lander (Addison-Wesley)
- *The Art of R Programming: A Tour of Statistical Software Design* by Norman Matloff (No Starch Press)
- *Python for Data Analysis* by Wes McKinney (O'Reilly)

*Data Analysis and Statistical Inference*
- *Statistical Inference* by George Casella and Roger L. Berger (Cengage Learning)
- *Bayesian Data Analysis* by Andrew Gelman, et al. (Chapman & Hall)
- *Data Analysis Using Regression and Multilevel/Hierarchical Models* by Andrew Gelman and Jennifer Hill (Cambridge University Press)
- *Advanced Data Analysis from an Elementary Point of View* (*http://goo.gl/udICRX*) by Cosma Shalizi (under contract with Cambridge University Press)
- *The Elements of Statistical Learning: Data Mining, Inference and Prediction* by Trevor Hastie, Robert Tibshirani, and Jerome Friedman (Springer)

*Artificial Intelligence and Machine Learning*
- *Pattern Recognition and Machine Learning* by Christopher Bishop (Springer)
- *Bayesian Reasoning and Machine Learning* by David Barber (Cambridge University Press)
- *Programming Collective Intelligence* by Toby Segaran (O'Reilly)
- *Artificial Intelligence: A Modern Approach* by Stuart Russell and Peter Norvig (Prentice Hall)

- *Foundations of Machine Learning* by Mehryar Mohri, Afshin Rostamizadeh, and Ameet Talwalkar (MIT Press)
- *Introduction to Machine Learning (Adaptive Computation and Machine Learning)* by Ethem Alpaydin (MIT Press)

*Experimental Design*
- *Field Experiments* by Alan S. Gerber and Donald P. Green (Norton)
- *Statistics for Experimenters: Design, Innovation, and Discovery* by George E. P. Box, et al. (Wiley-Interscience)

*Visualization*
- *The Elements of Graphing Data* by William Cleveland (Hobart Press)
- *Visualize This: The FlowingData Guide to Design, Visualization, and Statistics* by Nathan Yau (Wiley)
- *The Visual Display of Quantitative Information* by Edward Tufte (Graphics Press)

# About the Contributors

The course would not have been a success without the many guest lecturers that came to speak to the class. While I gave some of the lectures, a large majority were given by guests from startups and tech companies, as well as professors from Columbia University. Most chapters in this book are based on those lectures. While generally speaking the contributors did not write the book, they contributed many of the ideas and content of the book, reviewed their chapters and offered feedback, and we're grateful to them. The class and book would not have existed without them. I invited them to speak in the class because I hold them up as role models for aspiring data scientists.

# Conventions Used in This Book

The following typographical conventions are used in this book:

*Italic*
Indicates new terms, URLs, email addresses, filenames, and file extensions.

*Constant width*

Used for program listings, as well as within paragraphs to refer to program elements such as variable or function names, databases, data types, environment variables, statements, and keywords.

**Constant width bold**

Shows commands or other text that should be typed literally by the user.

*Constant width italic*

Shows text that should be replaced with user-supplied values or by values determined by context.

 This icon signifies a tip, suggestion, or general note.

 This icon indicates a warning or caution.

# Using Code Examples

Supplemental material (datasets, exercises, etc.) is available for download at *https://github.com/oreillymedia/doing_data_science*.

This book is here to help you get your job done. In general, if example code is offered with this book, you may use it in your programs and documentation. You do not need to contact us for permission unless you're reproducing a significant portion of the code. For example, writing a program that uses several chunks of code from this book does not require permission. Selling or distributing a CD-ROM of examples from O'Reilly books does require permission. Answering a question by citing this book and quoting example code does not require permission. Incorporating a significant amount of example code from this book into your product's documentation does require permission.

We appreciate, but do not require, attribution. An attribution usually includes the title, author, publisher, and ISBN. For example: *"Doing*

*Data Science* by Cathy O'Neil and Rachel Schutt (O'Reilly). Copyright 2014 Cathy O'Neil and Rachel Schutt, 978-1-449-35865-5."

If you feel your use of code examples falls outside fair use or the permission given above, feel free to contact us at *permissions@oreilly.com*.

## Safari® Books Online

*Safari Books Online* is an on-demand digital library that delivers expert content in both book and video form from the world's leading authors in technology and business.

Technology professionals, software developers, web designers, and business and creative professionals use Safari Books Online as their primary resource for research, problem solving, learning, and certification training.

Safari Books Online offers a range of plans and pricing for enterprise, government, education, and individuals.

Members have access to thousands of books, training videos, and prepublication manuscripts in one fully searchable database from publishers like O'Reilly Media, Prentice Hall Professional, Addison-Wesley Professional, Microsoft Press, Sams, Que, Peachpit Press, Focal Press, Cisco Press, John Wiley & Sons, Syngress, Morgan Kaufmann, IBM Redbooks, Packt, Adobe Press, FT Press, Apress, Manning, New Riders, McGraw-Hill, Jones & Bartlett, Course Technology, and hundreds more. For more information about Safari Books Online, please visit us online.

## How to Contact Us

Please address comments and questions concerning this book to the publisher:

O'Reilly Media, Inc.
1005 Gravenstein Highway North
Sebastopol, CA 95472
800-998-9938 (in the United States or Canada)
707-829-0515 (international or local)
707-829-0104 (fax)

We have a web page for this book, where we list errata, examples, and any additional information. You can access this page at *http://oreil.ly/ doing_data_science*.

To comment or ask technical questions about this book, send email to *bookquestions@oreilly.com*.

For more information about our books, courses, conferences, and news, see our website at *http://www.oreilly.com*.

Find us on Facebook: *http://facebook.com/oreilly*

Follow us on Twitter: *http://twitter.com/oreillymedia*

Watch us on YouTube: *http://www.youtube.com/oreillymedia*

## Acknowledgments

Rachel would like to thank her Google influences: David Huffaker, Makoto Uchida, Andrew Tomkins, Abhijit Bose, Daryl Pregibon, Diane Lambert, Josh Wills, David Crawshaw, David Gibson, Corinna Cortes, Zach Yeskel, and Gueorgi Kossinetts. From the Columbia statistics department: Andrew Gelman and David Madigan; and the lab instructor and teaching assistant for the course, Jared Lander and Ben Reddy.

Rachel appreciates the loving support of family and friends, especially Eran Goldshtein, Barbara and Schutt, Becky, Susie and Alex, Nick, Lilah, Belle, Shahed, and the Feeneys.

Cathy would like to thank her family and friends, including her wonderful sons and husband, who let her go off once a week to blog the evening class.

We both would like to thank:

- The brain trust that convened in Cathy's apartment: Chris Wiggins, David Madigan, Mark Hansen, Jake Hofman, Ori Stitelman, and Brian Dalessandro.
- Our editors, Courtney Nash and Mike Loukides.
- The participants and organizers of the IMA User-level modeling conference where some preliminary conversations took place.
- The students!

- Coppelia, where Cathy and Rachel met for breakfast a lot.

We'd also like to thank John Johnson and David Park of Johnson Research Labs for their generosity and the luxury of time to spend writing this book.

# CHAPTER 1

# Introduction: What Is Data Science?

Over the past few years, there's been a lot of hype in the media about "data science" and "Big Data." A reasonable first reaction to all of this might be some combination of skepticism and confusion; indeed we, Cathy and Rachel, had that exact reaction.

And we let ourselves indulge in our bewilderment for a while, first separately, and then, once we met, together over many Wednesday morning breakfasts. But we couldn't get rid of a nagging feeling that there was something *real* there, perhaps something deep and profound representing a paradigm shift in our culture around data. Perhaps, we considered, it's even a paradigm shift that plays to our strengths. Instead of ignoring it, we decided to explore it more.

But before we go into that, let's first delve into what struck us as confusing and vague—perhaps you've had similar inclinations. After that we'll explain what made us get past our own concerns, to the point where Rachel created a course on data science at Columbia University, Cathy blogged the course, and you're now reading a book based on it.

## Big Data and Data Science Hype

Let's get this out of the way right off the bat, because many of you are likely skeptical of data science already for many of the reasons we were. We want to address this up front to let you know: *we're right there with you.* If you're a skeptic too, it probably means you have something useful to contribute to making data science into a more legitimate field that has the power to have a positive impact on society.

So, what is eyebrow-raising about Big Data and data science? Let's count the ways:

1. There's a lack of definitions around the most basic terminology. What is "Big Data" anyway? What does "data science" mean? What is the relationship between Big Data and data science? Is data science the science of Big Data? Is data science only the stuff going on in companies like Google and Facebook and tech companies? Why do many people refer to Big Data as crossing disciplines (astronomy, finance, tech, etc.) and to data science as only taking place in tech? Just how *big* is big? Or is it just a relative term? These terms are so ambiguous, they're well-nigh meaningless.

2. There's a distinct lack of respect for the researchers in academia and industry labs who have been working on this kind of stuff for years, and whose work is based on decades (in some cases, centuries) of work by statisticians, computer scientists, mathematicians, engineers, and scientists of all types. From the way the media describes it, machine learning algorithms were just invented last week and data was never "big" until Google came along. This is simply not the case. Many of the methods and techniques we're using—and the challenges we're facing now—are part of the evolution of everything that's come before. This doesn't mean that there's not new and exciting stuff going on, but we think it's important to show some basic respect for everything that came before.

3. The hype is crazy—people throw around tired phrases straight out of the height of the pre-financial crisis era like "Masters of the Universe" to describe data scientists, and that doesn't bode well. In general, hype masks reality and increases the noise-to-signal ratio. The longer the hype goes on, the more many of us will get turned off by it, and the harder it will be to see what's good underneath it all, if anything.

4. Statisticians already feel that they are studying and working on the "Science of Data." That's their bread and butter. Maybe you, dear reader, are not a statistican and don't care, but imagine that for the statistician, this feels a little bit like how identity theft might feel for you. Although we will make the case that data science is *not* just a rebranding of statistics or machine learning but rather

a field unto itself, the media often describes data science in a way that makes it sound like as if it's simply statistics or machine learning in the context of the tech industry.

5. People have said to us, "Anything that has to call itself a science isn't." Although there might be truth in there, that doesn't mean that the term "data science" *itself* represents nothing, but of course what it represents may not be science but more of a craft.

# Getting Past the Hype

Rachel's experience going from getting a PhD in statistics to working at Google is a great example to illustrate why we thought, in spite of the aforementioned reasons to be dubious, there might be some meat in the data science sandwich. In her words:

> It was clear to me pretty quickly that the stuff I was working on at Google was different than anything I had learned at school when I got my PhD in statistics. This is not to say that my degree was useless; far from it—what I'd learned in school provided a framework and way of thinking that I relied on daily, and much of the actual content provided a solid theoretical and practical foundation necessary to do my work.

> But there were also many skills I had to acquire on the job at Google that I *hadn't* learned in school. Of course, my experience is specific to me in the sense that I had a statistics background and picked up more computation, coding, and visualization skills, as well as domain expertise while at Google. Another person coming in as a computer scientist or a social scientist or a physicist would have different gaps and would fill them in accordingly. But what is important here is that, as individuals, we each had different strengths and gaps, yet we were able to solve problems by putting ourselves together into a data team well-suited to solve the data problems that came our way.

Here's a reasonable response you might have to this story. It's a general truism that, whenever you go from school to a real job, you realize there's a gap between what you learned in school and what you do on the job. In other words, you were simply facing the difference between academic statistics and industry statistics.

We have a couple replies to this:

- Sure, there's is a difference between industry and academia. But does it really have to be that way? Why do many courses in school have to be so intrinsically out of touch with reality?

- Even so, the gap doesn't represent simply a difference between industry statistics and academic statistics. The general experience of data scientists is that, at their job, they have access to a *larger body of knowledge and methodology*, as well as a process, which we now define as the *data science process* (details in Chapter 2), that has foundations in both statistics and computer science.

Around all the hype, in other words, there is a ring of truth: this *is* something new. But at the same time, it's a fragile, nascent idea at real risk of being rejected prematurely. For one thing, it's being paraded around as a magic bullet, raising unrealistic expectations that will surely be disappointed.

Rachel gave herself the task of understanding the cultural phenomenon of data science and how others were experiencing it. She started meeting with people at Google, at startups and tech companies, and at universities, mostly from within statistics departments.

From those meetings she started to form a clearer picture of the new thing that's emerging. She ultimately decided to continue the investigation by giving a course at Columbia called "Introduction to Data Science," which Cathy covered on her blog. We figured that by the end of the semester, we, and hopefully the students, would know what all this actually meant. And now, with this book, we hope to do the same for many more people.

# Why Now?

We have massive amounts of data about many aspects of our lives, and, simultaneously, an abundance of inexpensive computing power. Shopping, communicating, reading news, listening to music, searching for information, expressing our opinions—all this is being tracked online, as most people know.

What people might not know is that the "datafication" of our offline behavior has started as well, mirroring the online data collection revolution (more on this later). Put the two together, and there's a lot to learn about our behavior and, by extension, who we are as a species.

It's not just Internet data, though—it's finance, the medical industry, pharmaceuticals, bioinformatics, social welfare, government, education, retail, and the list goes on. There is a growing influence of data in most sectors and most industries. In some cases, the amount of data collected might be enough to be considered "big" (more on this in the next chapter); in other cases, it's not.

But it's not only the massiveness that makes all this new data interesting (or poses challenges). It's that the data itself, often in real time, becomes the building blocks of data *products*. On the Internet, this means Amazon recommendation systems, friend recommendations on Facebook, film and music recommendations, and so on. In finance, this means credit ratings, trading algorithms, and models. In education, this is starting to mean dynamic personalized learning and assessments coming out of places like Knewton and Khan Academy. In government, this means policies based on data.

We're witnessing the beginning of a massive, culturally saturated feedback loop where our behavior changes the product and the product changes our behavior. Technology makes this possible: infrastructure for large-scale data processing, increased memory, and bandwidth, as well as a cultural acceptance of technology in the fabric of our lives. This wasn't true a decade ago.

Considering the impact of this feedback loop, we should start thinking seriously about how it's being conducted, along with the ethical and technical responsibilities for the people responsible for the process. One goal of this book is a first stab at that conversation.

## Datafication

In the May/June 2013 issue of *Foreign Affairs*, Kenneth Neil Cukier and Viktor Mayer-Schoenberger wrote an article called "The Rise of Big Data". In it they discuss the concept of datafication, and their example is how we quantify friendships with "likes": it's the way everything we do, online or otherwise, ends up recorded for later examination in someone's data storage units. Or maybe multiple storage units, and maybe also for sale.

They define datafication as a process of "taking all aspects of life and turning them into data." As examples, they mention that "Google's augmented-reality glasses datafy the gaze. Twitter datafies stray thoughts. LinkedIn datafies professional networks."

Datafication is an interesting concept and led us to consider its importance with respect to people's intentions about sharing their own data. We are being datafied, or rather our actions are, and when we "like" someone or something online, we are intending to be datafied, or at least we should expect to be. But when we merely browse the Web, we are unintentionally, or at least passively, being datafied through cookies that we might or might not be aware of. And when we walk around in a store, or even on the street, we are being datafied in a completely unintentional way, via sensors, cameras, or Google glasses.

This spectrum of intentionality ranges from us gleefully taking part in a social media experiment we are proud of, to all-out surveillance and stalking. But it's all datafication. Our intentions may run the gamut, but the results don't.

They follow up their definition in the article with a line that speaks volumes about their perspective:

> Once we datafy things, we can transform their purpose and turn the information into new forms of value.

Here's an important question that we will come back to throughout the book: who is "we" in that case? What kinds of *value* do they refer to? Mostly, given their examples, the "we" is the modelers and entrepreneurs making money from getting people to buy stuff, and the "value" translates into something like increased efficiency through automation.

If we want to think bigger, if we want our "we" to refer to people in general, we'll be swimming against the tide.

# The Current Landscape (with a Little History)

So, what is data science? Is it new, or is it just statistics or analytics rebranded? Is it real, or is it pure hype? And if it's new and if it's real, what does that mean?

This is an ongoing discussion, but one way to understand what's going on in this industry is to look online and see what current discussions are taking place. This doesn't necessarily tell us what data science is, but it at least tells us what other people think it is, or how they're perceiving it. For example, on Quora there's a discussion from 2010 about "What is Data Science?" and here's Metamarket CEO Mike Driscoll's answer:

Data science, as it's practiced, is a blend of Red-Bull-fueled hacking and espresso-inspired statistics.

But data science is not merely hacking—because when hackers finish debugging their Bash one-liners and Pig scripts, few of them care about non-Euclidean distance metrics.

And data science is not merely statistics, because when statisticians finish theorizing the perfect model, few could read a tab-delimited file into R if their job depended on it.

Data science is the civil engineering of data. Its acolytes possess a practical knowledge of tools and materials, coupled with a theoretical understanding of what's possible.

Driscoll then refers to Drew Conway's Venn diagram of data science from 2010, shown in Figure 1-1.

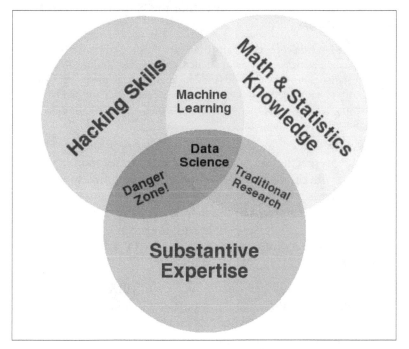

*Figure 1-1. Drew Conway's Venn diagram of data science*

He also mentions the sexy skills of data geeks from Nathan Yau's 2009 post, "Rise of the Data Scientist", which include:

- Statistics (traditional analysis you're used to thinking about)
- Data munging (parsing, scraping, and formatting data)

- Visualization (graphs, tools, etc.)

But wait, is data science just a bag of tricks? Or is it the logical extension of other fields like statistics and machine learning?

For one argument, see Cosma Shalizi's posts here (*http://goo.gl/SO7ceN*) and here (*http://goo.gl/pXg1fU*), and Cathy's posts here (*http://goo.gl/F4K4hE*) and here (*http://goo.gl/X9Bmxj*), which constitute an ongoing discussion of the difference between a statistician and a data scientist. Cosma basically argues that any statistics department worth its salt does all the stuff in the descriptions of data science that he sees, and therefore data science is just a rebranding and unwelcome takeover of statistics.

For a slightly different perspective, see ASA President Nancy Geller's 2011 Amstat News article, "Don't shun the 'S' word", in which she defends statistics:

> We need to tell people that Statisticians are the ones who make sense of the data deluge occurring in science, engineering, and medicine; that statistics provides methods for data analysis in all fields, from art history to zoology; that it is exciting to be a Statistician in the 21st century because of the many challenges brought about by the data explosion in all of these fields.

Though we get her point—the phrase "art history to zoology" is supposed to represent the concept of A to Z—she's kind of shooting herself in the foot with these examples because they don't correspond to the high-tech world where much of the data explosion is coming from. Much of the development of the field is happening in industry, not academia. That is, there are people with the job title data scientist in companies, but no professors of data science in academia. (Though this may be changing.)

Not long ago, DJ Patil described how he and Jeff Hammerbacher—then at LinkedIn and Facebook, respectively—coined the term "data scientist" in 2008. So that is when "data scientist" emerged as a job title. (Wikipedia finally gained an entry on data science in 2012.)

It makes sense to us that once the skill set required to thrive at Google —working with a team on problems that required a hybrid skill set of stats and computer science paired with personal characteristics including curiosity and persistence—spread to other Silicon Valley tech companies, it required a new job title. Once it became a pattern, it deserved a name. And once it got a name, everyone and their mother wanted to be one. It got even worse when *Harvard Business Review* declared data scientist to be the "Sexiest Job of the 21st Century".

## The Role of the Social Scientist in Data Science

Both LinkedIn and Facebook are social network companies. Oftentimes a description or definition of data scientist includes hybrid statistician, software engineer, and social scientist. This made sense in the context of companies where the product was a *social* product and still makes sense when we're dealing with human or user behavior. But if you think about Drew Conway's Venn diagram, data science problems cross disciplines—that's what the substantive expertise is referring to.

In other words, it depends on the context of the problems you're trying to solve. If they're social science-y problems like friend recommendations or people you know or user segmentation, then by all means, bring on the social scientist! Social scientists also do tend to be good question askers and have other good investigative qualities, so a social scientist who also has the quantitative and programming chops makes a great data scientist.

But it's almost a "historical" (historical is in quotes because 2008 isn't that long ago) artifact to limit your conception of a data scientist to someone who works only with online user behavior data. There's another emerging field out there called computational social sciences, which could be thought of as a subset of data science.

But we can go back even further. In 2001, William Cleveland wrote a position paper about data science called "Data Science: An action plan to expand the field of statistics."

So data science existed before data scientists? Is this semantics, or does it make sense?

This all begs a few questions: can you define data science by what data scientists *do*? Who gets to define the field, anyway? There's lots of buzz and hype—does the media get to define it, or should we rely on the practitioners, the self-appointed data scientists? Or is there some actual authority? Let's leave these as open questions for now, though we will return to them throughout the book.

## Data Science Jobs

Columbia just decided to start an Institute for Data Sciences and Engineering with Bloomberg's help. There are 465 job openings in New York City alone for data scientists last time we checked. That's a lot. So even if data science isn't a real field, it has *real* jobs.

And here's one thing we noticed about most of the job descriptions: they ask data scientists to be experts in computer science, statistics, communication, data visualization, *and* to have extensive domain expertise. Nobody is an expert in everything, which is why it makes more sense to create teams of people who have different profiles and different expertise—together, as a team, they can specialize in all those things. We'll talk about this more after we look at the composite set of skills in demand for today's data scientists.

# A Data Science Profile

In the class, Rachel handed out index cards and asked everyone to profile themselves (on a relative rather than absolute scale) with respect to their skill levels in the following domains:

- Computer science
- Math
- Statistics
- Machine learning
- Domain expertise
- Communication and presentation skills
- Data visualization

As an example, Figure 1-2 shows Rachel's data science profile.

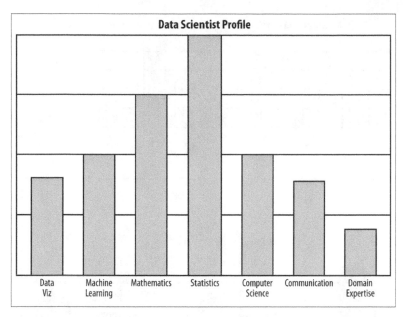

*Figure 1-2. Rachel's data science profile, which she created to illustrate trying to visualize oneself as a data scientist; she wanted students and guest lecturers to "riff" on this—to add buckets or remove skills, use a different scale or visualization method, and think about the drawbacks of self-reporting*

We taped the index cards to the blackboard and got to see how everyone else thought of themselves. There was quite a bit of variation, which is cool—lots of people in the class were coming from social sciences, for example.

Where is your data science profile at the moment, and where would you like it to be in a few months, or years?

As we mentioned earlier, a data science team works best when different skills (profiles) are represented across different people, because nobody is good at everything. It makes us wonder if it might be more worthwhile to define a "data science team"—as shown in Figure 1-3—than to define a data scientist.

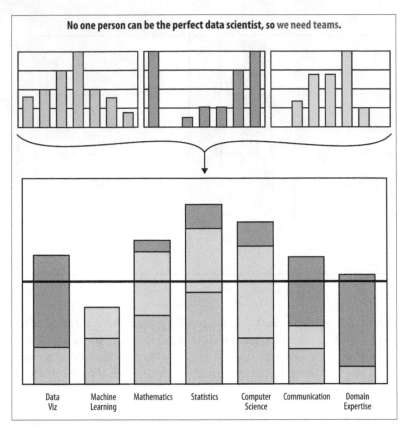

*Figure 1-3. Data science team profiles can be constructed from data scientist profiles; there should be alignment between the data science team profile and the profile of the data problems they try to solve*

# Thought Experiment: Meta-Definition

Every class had at least one thought experiment that the students discussed in groups. Most of the thought experiments were very open-ended, and the intention was to provoke discussion about a wide variety of topics related to data science. For the first class, the initial thought experiment was: *can we use data science to define data science?*

The class broke into small groups to think about and discuss this question. Here are a few interesting things that emerged from those conversations:

*Start with a text-mining model.*

We could do a Google search for "data science" and perform a text-mining model. But that would depend on us being a *usagist* rather than a *prescriptionist* with respect to language. A usagist would let the masses define data science (where "the masses" refers to whatever Google's search engine finds). Would it be better to be a prescriptionist and refer to an authority such as the *Oxford English Dictionary*? Unfortunately, the *OED* probably doesn't have an entry yet, and we don't have time to wait for it. Let's agree that there's a spectrum, that one authority doesn't feel right, and that "the masses" doesn't either.

*So what about a clustering algorithm?*

How about we look at practitioners of data science and see how *they* describe what they do (maybe in a word cloud for starters)? Then we can look at how people who claim to be other things like statisticians or physicists or economists describe what they do. From there, we can try to use a clustering algorithm (which we'll use in Chapter 3) or some other model and see if, when it gets as input "the stuff someone does," it gives a good prediction on what field that person is in.

Just for comparison, check out what Harlan Harris recently did related to the field of data science: he took a survey and used clustering to define subfields of data science, which gave rise to Figure 1-4.

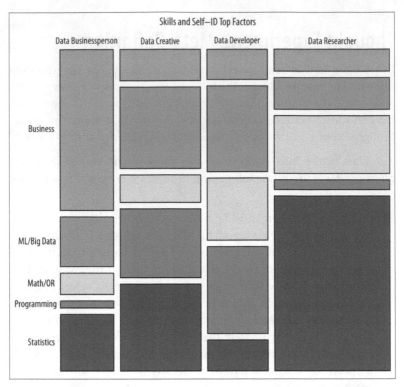

*Figure 1-4. Harlan Harris's clustering and visualization of subfields of data science from Analyzing the Analyzers (O'Reilly) by Harlan Harris, Sean Murphy, and Marck Vaisman based on a survey of several hundred data science practitioners in mid-2012*

# OK, So What Is a Data Scientist, Really?

Perhaps the most concrete approach is to define data science is by its usage—e.g., what data scientists get paid to do. With that as motivation, we'll describe what data scientists do. And we'll cheat a bit by talking first about data scientists in academia.

## In Academia

The reality is that currently, no one calls themselves a data scientist in academia, except to take on a secondary title for the sake of being a part of a "data science institute" at a university, or for applying for a grant that supplies money for data science research.

Instead, let's ask a related question: who in academia plans to *become* a data scientist? There were 60 students in the Intro to Data Science class at Columbia. When Rachel proposed the course, she assumed the makeup of the students would mainly be statisticians, applied mathematicians, and computer scientists. Actually, though, it ended up being those people plus sociologists, journalists, political scientists, biomedical informatics students, students from NYC government agencies and nonprofits related to social welfare, someone from the architecture school, others from environmental engineering, pure mathematicians, business marketing students, and students who already worked as data scientists. They were all interested in figuring out ways to solve important problems, often of social value, with data.

For the term "data science" to catch on in academia at the level of the faculty, and as a primary title, the research area needs to be more formally defined. Note there is already a rich set of problems that could translate into many PhD theses.

Here's a stab at what this could look like: an academic data scientist is a scientist, trained in anything from social science to biology, who works with large amounts of data, and must grapple with computational problems posed by the structure, size, messiness, and the complexity and nature of the data, while simultaneously solving a real-world problem.

The case for articulating it like this is as follows: across academic disciplines, the computational and deep data problems have major commonalities. If researchers across departments join forces, they can solve multiple real-world problems from different domains.

## In Industry

What do data scientists look like in industry? It depends on the level of seniority and whether you're talking about the Internet/online industry in particular. The role of data scientist need not be exclusive to the tech world, but that's where the term originated; so for the purposes of the conversation, let us say what it means there.

A chief data scientist should be setting the data strategy of the company, which involves a variety of things: setting everything up from the engineering and infrastructure for collecting data and logging, to privacy concerns, to deciding what data will be user-facing, how data is going to be used to make decisions, and how it's going to be built back into the product. She should manage a team of engineers, scientists, and analysts and should communicate with leadership across the company, including the CEO, CTO, and product leadership. She'll also be concerned with patenting innovative solutions and setting research goals.

More generally, a data scientist is someone who knows how to extract meaning from and interpret data, which requires both tools and methods from statistics and machine learning, as well as being human. She spends a lot of time in the process of collecting, cleaning, and munging data, because data is never clean. This process requires persistence, statistics, and software engineering skills—skills that are also necessary for understanding biases in the data, and for debugging logging output from code.

Once she gets the data into shape, a crucial part is exploratory data analysis, which combines visualization and data sense. She'll find patterns, build models, and algorithms—some with the intention of understanding product usage and the overall health of the product, and others to serve as prototypes that ultimately get baked back into the product. She may design experiments, and she is a critical part of data-driven decision making. She'll communicate with team members, engineers, and leadership in clear language and with data visualizations so that even if her colleagues are not immersed in the data themselves, they will understand the implications.

That's the high-level picture, and this book is about helping you understand the vast majority of it. We're done with *talking* about data science; let's go ahead and *do* some!

# CHAPTER 2

# Statistical Inference, Exploratory Data Analysis, and the Data Science Process

We begin this chapter with a discussion of statistical inference and statistical thinking. Next we explore what we feel every data scientist should do once they've gotten data in hand for any data-related project: exploratory data analysis (EDA).

From there, we move into looking at what we're defining as the data science process in a little more detail. We'll end with a thought experiment and a case study.

## Statistical Thinking in the Age of Big Data

> Big Data is a vague term, used loosely, if often, these days. But put simply, the catchall phrase means three things. First, it is a bundle of technologies. Second, it is a potential revolution in measurement. And third, it is a point of view, or philosophy, about how decisions will be—and perhaps should be—made in the future.
>
> — Steve Lohr
> *The New York Times*

When you're developing your skill set as a data scientist, certain foundational pieces need to be in place first—statistics, linear algebra, some programming. Even once you have those pieces, part of the challenge is that you will be developing several skill sets in parallel simultaneously—data preparation and munging, modeling, coding, visualization, and communication—that are interdependent. As we progress through the book, these threads will be intertwined. That said, we need to start somewhere, and will begin by getting grounded in statistical inference.

We expect the readers of this book to have diverse backgrounds. For example, some of you might already be awesome software engineers who can build data pipelines and code with the best of them but don't know much about statistics; others might be marketing analysts who don't really know how to code at all yet; and others might be curious, smart people who want to know what this data science thing is all about.

So while we're asking that readers already have certain prerequisites down, we can't come to your house and look at your transcript to make sure you actually have taken a statistics course, or have read a statistics book before. And even if you have taken Introduction to Statistics—a course we know from many awkward cocktail party conversations that 99% of people dreaded and wish they'd never had to take—this likely gave you no flavor for the depth and beauty of statistical inference.

But even if it did, and maybe you're a PhD-level statistician, it's always helpful to go back to fundamentals and remind ourselves of what statistical inference and thinking is all about. And further still, in the age of Big Data, classical statistics methods need to be revisited and reimagined in new contexts.

## Statistical Inference

The world we live in is complex, random, and uncertain. At the same time, it's one big data-generating machine.

As we commute to work on subways and in cars, as our blood moves through our bodies, as we're shopping, emailing, procrastinating at work by browsing the Internet and watching the stock market, as we're building things, eating things, talking to our friends and family about things, while factories are producing products, this all at least potentially produces data.

Imagine spending 24 hours looking out the window, and for every minute, counting and recording the number of people who pass by. Or gathering up everyone who lives within a mile of your house and making them tell you how many email messages they receive every day for the next year. Imagine heading over to your local hospital and rummaging around in the blood samples looking for patterns in the

DNA. That all sounded creepy, but it wasn't supposed to. The point here is that the processes in our lives are actually data-generating processes.

We'd like ways to describe, understand, and make sense of these processes, in part because as scientists we just want to understand the world better, but many times, understanding these processes is part of the solution to problems we're trying to solve.

Data represents the traces of the real-world processes, and exactly which traces we gather are decided by our data collection or sampling method. You, the data scientist, the observer, are turning the world into data, and this is an utterly subjective, not objective, process.

After separating the process from the data collection, we can see clearly that there are two sources of randomness and uncertainty. Namely, the randomness and uncertainty underlying the process itself, and the uncertainty associated with your underlying data collection methods.

Once you have all this data, you have somehow captured the world, or certain traces of the world. But you can't go walking around with a huge Excel spreadsheet or database of millions of transactions and look at it and, with a snap of a finger, understand the world and process that generated it.

So you need a new idea, and that's to simplify those captured traces into something more comprehensible, to something that somehow captures it all in a much more concise way, and that something could be mathematical models or functions of the data, known as statistical estimators.

This overall process of going from the world to the data, and then from the data back to the world, is the field of *statistical inference*.

More precisely, statistical inference is the discipline that concerns itself with the development of procedures, methods, and theorems that allow us to extract meaning and information from data that has been generated by stochastic (random) processes.

## Populations and Samples

Let's get some terminology and concepts in place to make sure we're all talking about the same thing.

In classical statistical literature, a distinction is made between the population and the sample. The word *population* immediately makes us think of the entire US population of 300 million people, or the entire world's population of 7 billion people. But put that image out of your head, because in statistical inference population isn't used to simply describe only people. It could be any set of objects or units, such as tweets or photographs or stars.

If we could measure the characteristics or extract characteristics of all those objects, we'd have a complete set of *observations*, and the convention is to use $N$ to represent the total number of observations in the population.

Suppose your population was all emails sent last year by employees at a huge corporation, BigCorp. Then a single observation could be a list of things: the sender's name, the list of recipients, date sent, text of email, number of characters in the email, number of sentences in the email, number of verbs in the email, and the length of time until first reply.

When we take a *sample*, we take a subset of the units of size $n$ in order to examine the observations to draw conclusions and make inferences about the population. There are different ways you might go about getting this subset of data, and you want to be aware of this sampling mechanism because it can introduce *biases* into the data, and distort it, so that the subset is not a "mini-me" shrunk-down version of the population. Once that happens, any conclusions you draw will simply be wrong and distorted.

In the BigCorp email example, you could make a list of all the employees and select 1/10th of those people *at random* and take all the email they ever sent, and that would be your sample. Alternatively, you could sample 1/10th of all email sent each day at random, and that would be your sample. Both these methods are reasonable, and both methods yield the same sample size. But if you took them and counted how many email messages each person sent, and used that to estimate the underlying *distribution* of emails sent by all indiviuals at BigCorp, you might get entirely different answers.

So if even getting a basic thing down like counting can get distorted when you're using a reasonable-sounding sampling method, imagine what can happen to more complicated algorithms and models if you haven't taken into account the process that got the data into your hands.

# Populations and Samples of Big Data

But, wait! In the age of Big Data, where we can record all users' actions all the time, don't we observe *everything*? Is there really still this notion of population and sample? If we had all the email in the first place, why would we need to take a sample?

With these questions, we've gotten to the heart of the matter. There are multiple aspects of this that need to be addressed.

*Sampling solves some engineering challenges*

In the current popular discussion of Big Data, the focus on enterprise solutions such as Hadoop to handle engineering and computational challenges caused by too much data overlooks sampling as a legitimate solution. At Google, for example, software engineers, data scientists, and statisticians sample all the time.

How much data you need at hand really depends on what your goal is: for analysis or inference purposes, you typically don't need to store all the data all the time. On the other hand, for serving purposes you might: in order to render the correct information in a UI for a user, you need to have all the information for that particular user, for example.

*Bias*

Even if we have access to all of Facebook's or Google's or Twitter's data corpus, any inferences we make from that data should not be extended to draw conclusions about humans beyond those sets of users, or even those users for any particular day.

Kate Crawford, a principal scientist at Microsoft Research, describes in her Strata talk, "Hidden Biases of Big Data," how if you analyzed tweets immediately before and after Hurricane Sandy, you would think that most people were supermarket shopping pre-Sandy and partying post-Sandy. However, most of those tweets came from New Yorkers. First of all, they're heavier Twitter users than, say, the coastal New Jerseyans, and second of all, the coastal New Jerseyans were worrying about other stuff like their house falling down and didn't have time to tweet.

In other words, you would think that Hurricane Sandy wasn't all that bad if you used tweet data to understand it. The only conclusion you can actually draw is that this is what Hurricane Sandy was like for the subset of Twitter users (who themselves are not representative of the general US population), whose situation was not so bad that they didn't have time to tweet.

Note, too, that in this case, if you didn't *have context* and know about Hurricane Sandy, you wouldn't know enough to interpret this data properly.

## Sampling

Let's rethink what the population and the sample are in various contexts.

In statistics we often model the relationship between a population and a sample with an underlying mathematical process. So we make simplifying *assumptions* about the underlying truth, the mathematical structure, and shape of the underlying generative process that created the data. We observe only one particular realization of that generative process, which is that sample.

So if we think of all the emails at BigCorp as the population, and if we randomly sample from that population by reading some but not all emails, then that sampling process would create one particular sample. However, if we resampled we'd get a different set of observations.

The uncertainty created by such a sampling process has a name: the *sampling distribution*. But like that 2010 movie *Inception* with Leonardo DiCaprio, where he's in a dream within a dream within a dream, it's possible to instead think of the complete corpus of emails at BigCorp as not the population but as a sample.

This set of emails (and here is where we're getting philosophical, but that's what this is all about) could actually be only one single realization from some larger *super-population*, and if the Great Coin Tosser in the sky had spun again that day, a different set of emails would have been observed.

In this interpretation, we treat this set of emails as a sample that we are using to make inferences about the underlying generative process that is the email writing habits of all the employees at BigCorp.

*New kinds of data*

Gone are the days when data is just a bunch of numbers and categorical variables. A strong data scientist needs to be versatile and comfortable with dealing a variety of types of data, including:

- Traditional: numerical, categorical, or binary
- Text: emails, tweets, *New York Times* articles (see Chapter 4 or Chapter 7)
- Records: user-level data, timestamped event data, json-formatted log files (see Chapter 6 or Chapter 8)
- Geo-based location data: briefly touched on in this chapter with NYC housing data
- Network (see Chapter 10)
- Sensor data (not covered in this book)
- Images (not covered in this book)

These new kinds of data require us to think more carefully about what sampling means in these contexts.

For example, with the firehose of real-time streaming data, if you analyze a Facebook user-level dataset for a week of activity that you aggregated from timestamped event logs, will any conclusions you draw from this dataset be relevant next week or next year?

How do you sample from a network and preserve the complex network structure?

Many of these questions represent open research questions for the statistical and computer science communities. This is the frontier! Given that some of these are open research problems, in practice, data scientists do the best they can, and often are inventing novel methods as part of their jobs.

## Terminology: Big Data

We've been throwing around "Big Data" quite a lot already and are guilty of barely defining it beyond raising some big questions in the previous chapter.

A few ways to think about Big Data:

**"Big" is a moving target**. Constructing a threshold for Big Data such as 1 petabyte is meaningless because it makes it sound absolute. Only when the size becomes a challenge is it worth referring to it as "Big." So it's a relative term referring to when the size of the data outstrips the state-of-the-art current computational solutions (in terms of memory, storage, complexity, and processing speed) available to handle it. So in the 1970s this meant something different than it does today.

**"Big" is when you can't fit it on one machine**. Different individuals and companies have different computational resources available to them, so for a single scientist data is big if she can't fit it on one machine because she has to learn a whole new host of tools and methods once that happens.

**Big Data is a cultural phenomenon**. It describes how much data is part of our lives, precipitated by accelerated advances in technology.

**The 4 Vs**: Volume, variety, velocity, and value. Many people are circulating this as a way to characterize Big Data. Take from it what you will.

# Big Data Can Mean Big Assumptions

In Chapter 1, we mentioned the Cukier and Mayer-Schoenberger article "The Rise of Big Data." In it, they argue that the Big Data revolution consists of three things:

- Collecting and using a lot of data rather than small samples
- Accepting messiness in your data
- Giving up on knowing the causes

They describe these steps in a rather grand fashion by claiming that Big Data doesn't need to understand cause given that the data is so enormous. It doesn't need to worry about sampling error because it is literally *keeping track of the truth*. The way the article frames this is by claiming that the new approach of Big Data is letting "N=ALL."

## Can N=ALL?

Here's the thing: it's pretty much never all. And we are very often missing the very things we should care about most.

So, for example, as this InfoWorld post (*http://goo.gl/LeYKDR*) explains, Internet surveillance will never really work, because the very clever and tech-savvy criminals that we most want to catch are the very ones we will never be able to catch, because they're always a step ahead.

An example from that very article—election night polls—is in itself a great counter-example: even if we poll absolutely everyone who leaves the polling stations, we still don't count people who decided not to vote in the first place. And those might be the very people we'd need to talk to to understand our country's voting problems.

Indeed, we'd argue that the assumption we make that N=ALL is one of the biggest problems we face in the age of Big Data. It is, above all, a way of excluding the voices of people who don't have the time, energy, or access to cast their vote in all sorts of informal, possibly unannounced, elections.

Those people, busy working two jobs and spending time waiting for buses, become invisible when we tally up the votes without them. To you this might just mean that the recommendations you receive on Netflix don't seem very good because most of the people who bother to rate things on Netflix are young and might have different tastes than you, which skews the recommendation engine toward them. But there are plenty of much more insidious consequences stemming from this basic idea.

## Data is not objective

Another way in which the assumption that N=ALL can matter is that it often gets translated into the idea that data is *objective*. It is wrong to believe either that data is objective or that "data speaks," and beware of people who say otherwise.

We were recently reminded of it in a terrifying way by this New York Times article (*http://nyti.ms/18OEq6j*) on Big Data and recruiter hiring practices. At one point, a data scientist is quoted as saying, "Let's put everything in and let the data speak for itself."

If you read the whole article, you'll learn that this algorithm tries to find "diamond in the rough" types of people to hire. A worthy effort, but one that you have to think through.

Say you decided to compare women and men with the exact same qualifications that have been hired in the past, but then, looking into what happened next you learn that those women have tended to leave more often, get promoted less often, and give more negative feedback on their environments when compared to the men.

Your model might be likely to hire the man over the woman next time the two similar candidates showed up, rather than looking into the possibility that the company doesn't treat female employees well.

In other words, ignoring causation can be a flaw, rather than a feature. Models that ignore causation can add to historical problems instead of addressing them (we'll explore this more in Chapter 11). And data doesn't speak for itself. Data is just a quantitative, pale echo of the events of our society.

---

## n = 1

At the other end of the spectrum from N=ALL, we have $n = 1$, by which we mean a sample size of 1. In the old days a sample size of 1 would be ridiculous; you would never want to draw inferences about an entire population by looking at a single individual. And don't worry, that's still ridiculous. But the concept of $n = 1$ takes on new meaning in the age of Big Data, where for a single person, we actually can record tons of information about them, and in fact we might even sample from all the events or actions they took (for example, phone calls or keystrokes) in order to make inferences about them. This is what user-level modeling is about.

---

## Modeling

In the next chapter, we'll look at how we build models from the data we collect, but first we want to discuss what we even mean by this term.

Rachel had a recent phone conversation with someone about a *modeling* workshop, and several minutes into it she realized the word "model" meant completely different things to them. He was using it to mean *data models*—the representation one is choosing to store one's data, which is the realm of database managers—whereas she was

---

talking about *statistical models*, which is what much of this book is about. One of Andrew Gelman's blog posts on modeling was recently tweeted by people in the fashion industry, but that's a different issue.

Even if you've used the terms *statistical model* or *mathematical model* for years, is it even clear to yourself and to the people you're talking to what you mean? What makes a model a *model*? Also, while we're asking fundamental questions like this, what's the difference between a statistical model and a machine learning algorithm?

Before we dive deeply into that, let's add a bit of context with this deliberately provocative *Wired* magazine piece, "The End of Theory: The Data Deluge Makes the Scientific Method Obsolete," published in 2008 by Chris Anderson, then editor-in-chief.

Anderson equates massive amounts of data to complete information and argues no models are necessary and "correlation is enough"; e.g., that in the context of massive amounts of data, "they [Google] don't have to settle for models at all."

Really? We don't think so, and we don't think you'll think so either by the end of the book. But the sentiment is similar to the Cukier and Mayer-Schoenberger article we just discussed about N=ALL, so you might already be getting a sense of the profound confusion we're witnessing all around us.

To their credit, it's the press that's currently raising awareness of these questions and issues, and someone has to do it. Even so, it's hard to take when the opinion makers are people who don't actually work with data. Think critically about whether you buy what Anderson is saying; where you agree, disagree, or where you need more information to form an opinion.

Given that this is how the popular press is currently describing and influencing public perception of data science and modeling, it's incumbent upon us as data scientists to be aware of it and to chime in with informed comments.

With that context, then, what do we mean when we say *models*? And how do we use them as data scientists? To get at these questions, let's dive in.

## What is a model?

Humans try to understand the world around them by representing it in different ways. Architects capture attributes of buildings through blueprints and three-dimensional, scaled-down versions. Molecular biologists capture protein structure with three-dimensional visualizations of the connections between amino acids. Statisticians and data scientists capture the uncertainty and randomness of data-generating processes with mathematical functions that express the shape and structure of the data itself.

A model is our attempt to understand and represent the nature of reality through a particular lens, be it architectural, biological, or mathematical.

A model is an artificial construction where all extraneous detail has been removed or abstracted. Attention must always be paid to these abstracted details after a model has been analyzed to see what might have been overlooked.

In the case of proteins, a model of the protein backbone with side-chains by itself is removed from the laws of quantum mechanics that govern the behavior of the electrons, which ultimately dictate the structure and actions of proteins. In the case of a statistical model, we may have mistakenly excluded key variables, included irrelevant ones, or assumed a mathematical structure divorced from reality.

### Statistical modeling

Before you get too involved with the data and start coding, it's useful to draw a picture of what you think the underlying process might be with your model. What comes first? What influences what? What causes what? What's a test of that?

But different people think in different ways. Some prefer to express these kinds of relationships in terms of math. The mathematical expressions will be general enough that they have to include parameters, but the values of these parameters are not yet known.

In mathematical expressions, the convention is to use Greek letters for parameters and Latin letters for data. So, for example, if you have two columns of data, $x$ and $y$, and you think there's a linear relationship, you'd write down $y = \beta_0 + \beta_1 x$. You don't know what $\beta_0$ and $\beta_1$ are in terms of actual numbers yet, so they're the parameters.

Other people prefer pictures and will first draw a diagram of data flow, possibly with arrows, showing how things affect other things or what happens over time. This gives them an abstract picture of the relationships before choosing equations to express them.

---

## But how do you build a model?

How do you have any clue whatsoever what functional form the data should take? Truth is, it's part art and part science. And sadly, this is where you'll find the least guidance in textbooks, in spite of the fact that it's the key to the whole thing. After all, this is the part of the modeling process where you have to make a lot of assumptions about the underlying structure of reality, and we should have standards as to how we make those choices and how we explain them. But we don't have global standards, so we make them up as we go along, and hopefully in a thoughtful way.

We're admitting this here: where to start is not obvious. If it were, we'd know the meaning of life. However, we will do our best to demonstrate for you throughout the book how it's done.

One place to start is exploratory data analysis (EDA), which we will cover in a later section. This entails making plots and building intuition for your particular dataset. EDA helps out a lot, as well as trial and error and iteration.

To be honest, until you've done it a lot, it seems very mysterious. The best thing to do is start simply and then build in complexity. Do the dumbest thing you can think of first. It's probably not that dumb.

For example, you can (and should) plot histograms and look at scatterplots to start getting a feel for the data. Then you just try writing something down, even if it's wrong first (it will probably be wrong first, but that doesn't matter).

So try writing down a linear function (more on that in the next chapter). When you write it down, you force yourself to think: does this make *any* sense? If not, why? What would make *more sense*? You start simply and keep building it up in complexity, making assumptions, and writing your assumptions down. You can use full-blown sentences if it helps—e.g., "I assume that my users naturally cluster into about five groups because when I hear the sales rep talk about them, she has about five different types of people she talks about"—then taking your words and trying to express them as equations and code.

Remember, it's always good to start simply. There is a trade-off in modeling between simple and accurate. Simple models may be easier to interpret and understand. Oftentimes the crude, simple model gets you 90% of the way there and only takes a few hours to build and fit, whereas getting a more complex model might take months and only get you to 92%.

You'll start building up your arsenal of potential models throughout this book. Some of the building blocks of these models are *probability distributions*.

## Probability distributions

Probability distributions are the foundation of statistical models. When we get to linear regression and Naive Bayes, you will see how this happens in practice. One can take multiple semesters of courses on probability theory, and so it's a tall challenge to condense it down for you in a small section.

Back in the day, before computers, scientists observed real-world phenomena, took measurements, and noticed that certain mathematical shapes kept reappearing. The classical example is the height of humans, following a *normal* distribution—a bell-shaped curve, also called a Gaussian distribution, named after Gauss.

Other common shapes have been named after their observers as well (e.g., the Poisson distribution and the Weibull distribution), while other shapes such as Gamma distributions or exponential distributions are named after associated mathematical objects.

Natural processes tend to generate measurements whose empirical shape could be approximated by mathematical functions with a few parameters that could be estimated from the data.

Not *all* processes generate data that looks like a *named* distribution, but many do. We can use these functions as building blocks of our models. It's beyond the scope of the book to go into each of the distributions in detail, but we provide them in Figure 2-1 as an illustration of the various common shapes, and to remind you that they only have names because someone observed them enough times to think they deserved names. There is actually an infinite number of possible distributions.

They are to be interpreted as assigning a *probability* to a subset of possible outcomes, and have corresponding functions. For example, the normal distribution is written as:

$$N(x|\mu, \sigma) \sim \frac{1}{\sigma\sqrt{2\pi}} e^{-\frac{(x-\mu)^2}{2\sigma^2}}$$

The parameter $\mu$ is the mean and median and controls where the distribution is centered (because this is a symmetric distribution), and the parameter $\sigma$ controls how spread out the distribution is. This is the general functional form, but for specific real-world phenomenon, these parameters have actual numbers as values, which we can estimate from the data.

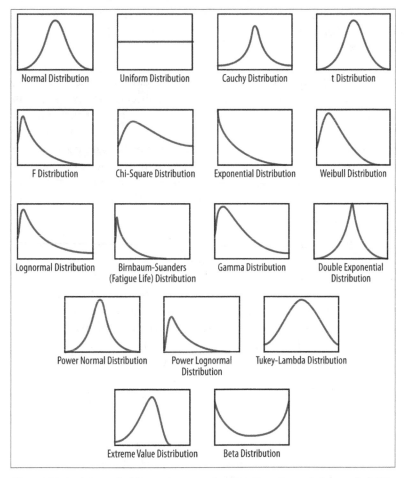

Figure 2-1. A bunch of continuous density functions (aka probability distributions)

A *random variable* denoted by $x$ or $y$ can be assumed to have a corresponding probability distribution, $p(x)$, which maps $x$ to a positive
real number. In order to be a probability density function, we're re-
stricted to the set of functions such that if we integrate $p(x)$ to get the
area under the curve, it is 1, so it can be interpreted as probability.

For example, let $x$ be the amount of time until the next bus arrives
(measured in minutes). $x$ is a random variable because there is variation and uncertainty in the amount of time until the next bus.

Suppose we know (for the sake of argument) that the time until the
next bus has a probability density function of $p(x) = 2e^{-2x}$. If we want
to know the likelihood of the next bus arriving in between 12 and 13
minutes, then we find the area under the curve between 12 and 13 by
$\int_{12}^{13} 2e^{-2x}$.

How do we know this is the right distribution to use? Well, there are
two possible ways: we can conduct an experiment where we show up
at the bus stop at a random time, measure how much time until the
next bus, and repeat this experiment over and over again. Then we
look at the measurements, plot them, and approximate the function
as discussed. Or, because we are familiar with the fact that "waiting
time" is a common enough real-world phenomenon that a distribution
called the exponential distribution has been invented to describe it, we
know that it takes the form $p(x) = \lambda e^{-\lambda x}$.

In addition to denoting distributions of single random variables with
functions of one variable, we use multivariate functions called *joint
distributions* to do the same thing for more than one random variable.
So in the case of two random variables, for example, we could denote
our distribution by a function $p(x, y)$, and it would take values in the
plane and give us nonnegative values. In keeping with its interpretation as a probability, its (double) integral over the whole plane would
be 1.

We also have what is called a *conditional distribution*, $p(x|y)$, which
is to be interpreted as the density function of $x$ given a particular value
of $y$.

When we're working with data, conditioning corresponds to subset-
ting. So for example, suppose we have a set of user-level data for
Amazon.com that lists for each user the amount of money spent last
month on Amazon, whether the user is male or female, and how many
items they looked at before adding the first item to the shopping cart.

If we consider $x$ to be the random variable that represents the amount of money spent, then we can look at the distribution of money spent across all users, and represent it as $p(x)$.

We can then take the subset of users who looked at more than five items before buying anything, and look at the distribution of money spent among these users. Let $y$ be the random variable that represents number of items looked at, then $p(x|y>5)$ would be the corresponding *conditional distribution*. Note a conditional distribution has the same properties as a regular distribution in that when we integrate it, it sums to 1 and has to take nonnegative values.

When we observe data points, i.e., $(x_1, y_1),(x_2, y_2),\ldots,(x_n, y_n)$, we are observing *realizations* of a pair of random variables. When we have an entire dataset with $n$ rows and $k$ columns, we are observing $n$ realizations of the joint distribution of those $k$ random variables.

For further reading on probability distributions, we recommend Sheldon Ross' book, *A First Course in Probability* (Pearson).

### Fitting a model

*Fitting* a model means that you estimate the parameters of the model using the observed data. You are using your data as evidence to help approximate the real-world mathematical process that generated the data. Fitting the model often involves optimization methods and algorithms, such as *maximum likelihood estimation*, to help get the parameters.

In fact, when you estimate the parameters, they are actually *estimators*, meaning they themselves are *functions* of the data. Once you fit the model, you actually can write it as $y = 7.2 + 4.5x$, for example, which means that your best guess is that this equation or functional form expresses the relationship between your two variables, based on your assumption that the data followed a linear pattern.

Fitting the model is when you start actually coding: your code will read in the data, and you'll specify the functional form that you wrote down on the piece of paper. Then R or Python will use built-in optimization methods to give you the most likely values of the parameters given the data.

As you gain sophistication, or if this is one of your areas of expertise, you'll dig around in the optimization methods yourself. Initially you should have an understanding that optimization is taking place and how it works, but you don't have to code this part yourself—it underlies the R or Python functions.

### Overfitting

Throughout the book you will be cautioned repeatedly about *overfitting*, possibly to the point you will have nightmares about it. Overfitting is the term used to mean that you used a dataset to estimate the parameters of your model, but your model isn't that good at capturing reality beyond your sampled data.

You might know this because you have tried to use it to predict labels for another set of data that you didn't use to fit the model, and it doesn't do a good job, as measured by an evaluation metric such as accuracy.

# Exploratory Data Analysis

> "Exploratory data analysis" is an attitude, a state of flexibility, a willingness to look for those things that we believe are not there, as well as those we believe to be there.
>
> — John Tukey

Earlier we mentioned exploratory data analysis (EDA) as the first step toward building a model. EDA is often relegated to chapter 1 (by which we mean the "easiest" and lowest level) of standard introductory statistics textbooks and then forgotten about for the rest of the book.

It's traditionally presented as a bunch of histograms and stem-and-leaf plots. They teach that stuff to kids in fifth grade so it seems trivial, right? No wonder no one thinks much of it.

But EDA is a critical part of the data science process, and also represents a philosophy or way of doing statistics practiced by a strain of statisticians coming from the Bell Labs tradition.

John Tukey, a mathematician at Bell Labs, developed exploratory data analysis in contrast to confirmatory data analysis, which concerns itself with modeling and hypotheses as described in the previous section. In EDA, there is no hypothesis and there is no model. The "exploratory" aspect means that your understanding of the problem you are solving, or might solve, is changing as you go.

## Historical Perspective: Bell Labs

Bell Labs is a research lab going back to the 1920s that has made innovations in physics, computer science, statistics, and math, producing languages like C++, and many Nobel Prize winners as well. There was a very successful and productive statistics group there, and among its many notable members was John Tukey, a mathematician who worked on a lot of statistical problems. He is considered the father of EDA and R (which started as the S language at Bell Labs; R is the open source version), and he was interested in trying to visualize high-dimensional data.

We think of Bell Labs as one of the places where data science was "born" because of the collaboration between disciplines, and the massive amounts of complex data available to people working there. It was a virtual playground for statisticians and computer scientists, much like Google is today.

In fact, in 2001, Bill Cleveland wrote "Data Science: An Action Plan for expanding the technical areas of the field of statistics," which described multidisciplinary investigation, models, and methods for data (traditional applied stats), computing with data (hardware, software, algorithms, coding), pedagogy, tool evaluation (staying on top of current trends in technology), and theory (the math behind the data).

You can read more about Bell Labs in the book *The Idea Factory* by Jon Gertner (Penguin Books).

The basic tools of EDA are plots, graphs and summary statistics. Generally speaking, it's a method of systematically going through the data, plotting distributions of all variables (using box plots), plotting time series of data, transforming variables, looking at all pairwise relationships between variables using scatterplot matrices, and generating summary statistics for all of them. At the very least that would mean computing their mean, minimum, maximum, the upper and lower quartiles, and identifying outliers.

But as much as EDA is a set of tools, it's also a mindset. And that mindset is about your relationship with the data. You want to understand the data—gain intuition, understand the shape of it, and try to connect your understanding of the process that generated the data to the data itself. EDA happens between you and the data and isn't about proving anything to anyone else yet.

## Philosophy of Exploratory Data Analysis

> Long before worrying about how to convince others, you first have to understand what's happening yourself.
>
> — Andrew Gelman

While at Google, Rachel was fortunate to work alongside two former Bell Labs/AT&T statisticians—Daryl Pregibon and Diane Lambert, who also work in this vein of applied statistics—and learned from them to make EDA a part of her best practices.

Yes, even with very large Google-scale data, they did EDA. In the context of data in an Internet/engineering company, EDA is done for some of the same reasons it's done with smaller datasets, but there are additional reasons to do it with data that has been generated from logs.

There are important reasons anyone working with data should do EDA. Namely, to gain intuition about the data; to make comparisons between distributions; for sanity checking (making sure the data is on the scale you expect, in the format you thought it should be); to find out where data is missing or if there are outliers; and to summarize the data.

In the context of data generated from logs, EDA also helps with debugging the logging process. For example, "patterns" you find in the data could actually be something wrong in the logging process that needs to be fixed. If you never go to the trouble of debugging, you'll continue to think your patterns are real. The engineers we've worked with are always grateful for help in this area.

In the end, EDA helps you make sure the product is performing as intended.

Although there's lots of visualization involved in EDA, we distinguish between EDA and data visualization in that EDA is done toward the beginning of analysis, and data visualization (which we'll get to in Chapter 9), as it's used in our vernacular, is done toward the end to communicate one's findings. With EDA, the graphics are solely done for *you* to understand what's going on.

With EDA, you can also use the understanding you get to inform and improve the development of algorithms. For example, suppose you are trying to develop a ranking algorithm that ranks content that you are showing to users. To do this you might want to develop a notion of "popular."

Before you decide how to quantify popularity (which could be, for example, highest frequency of clicks, or the post with the most number of comments, or comments above some threshold, or some weighted average of many metrics), you need to understand how the data is behaving, and the best way to do that is looking at it and getting your hands dirty.

Plotting data and making comparisons can get you extremely far, and is far better to do than getting a dataset and immediately running a regression just because you know how. It's been a disservice to analysts and data scientists that EDA has not been enforced as a critical part of the process of working with data. Take this opportunity to make it part of your process!

Here are some references to help you understand best practices and historical context:

1. *Exploratory Data Analysis* by John Tukey (Pearson)
2. *The Visual Display of Quantitative Information* by Edward Tufte (Graphics Press)
3. *The Elements of Graphing Data* by William S. Cleveland (Hobart Press)
4. *Statistical Graphics for Visualizing Multivariate Data* by William G. Jacoby (Sage)
5. "Exploratory Data Analysis for Complex Models" by Andrew Gelman (American Statistical Association)
6. *The Future of Data Analysis* by John Tukey. Annals of Mathematical Statistics, Volume 33, Number 1 (1962), 1-67.
7. *Data Analysis, Exploratory* by David Brillinger [8-page excerpt from *International Encyclopedia of Political Science* (Sage)]

## Exercise: EDA

There are 31 datasets named nyt1.csv, nyt2.csv,…,nyt31.csv, which you can find here: *https://github.com/oreillymedia/doing_data_science*.

Each one represents one (simulated) day's worth of ads shown and clicks recorded on the *New York Times* home page in May 2012. Each row represents a single user. There are five columns: age, gender (0=female, 1=male), number impressions, number clicks, and logged-in.

You'll be using R to handle these data. It's a programming language designed specifically for data analysis, and it's pretty intuitive to start using. You can download it here (*http://www.r-project.org*). Once you have it installed, you can load a single file into R with this command:

```
data1 <- read.csv(url("http://stat.columbia.edu/~rachel/
                            datasets/nyt1.csv"))
```

Once you have the data loaded, it's time for some EDA:

1. Create a new variable, `age_group`, that categorizes users as "<18", "18-24", "25-34", "35-44", "45-54", "55-64", and "65+".

2. For a single day:

   - Plot the distributions of number impressions and click-through-rate (CTR=# clicks/# impressions) for these six age categories.

   - Define a new variable to segment or categorize users based on their click behavior.

   - Explore the data and make visual and quantitative comparisons across user segments/demographics (<18-year-old males versus < 18-year-old females or logged-in versus not, for example).

   - Create metrics/measurements/statistics that summarize the data. Examples of potential metrics include CTR, quantiles, mean, median, variance, and max, and these can be calculated across the various user segments. Be selective. Think about what will be important to track over time—what will compress the data, but still capture user behavior.

3. Now extend your analysis across days. Visualize some metrics and distributions over time.

4. Describe and interpret any patterns you find.

## Sample code

Here we'll give you the beginning of a sample solution for this exercise. The reality is that we can't teach you about data science and teach you how to code all in the same book. Learning to code in a new language requires a lot of trial and error as well as going online and searching on Google or stackoverflow.

Chances are, if you're trying to figure out how to plot something or build a model in R, other people have tried as well, so rather than banging your head against the wall, look online. [Ed note: There might also be some books available to help you out on this front as well.] We suggest not looking at this code until you've struggled along a bit:

```
# Author: Maura Fitzgerald
data1 <- read.csv(url("http://stat.columbia.edu/~rachel/
                    datasets/nyt1.csv"))

# categorize
head(data1)
data1$agecat <-cut(data1$Age,c(-Inf,0,18,24,34,44,54,64,Inf))

# view
summary(data1)

# brackets
install.packages("doBy")
library("doBy")
siterange <- function(x){c(length(x), min(x), mean(x), max(x))}
summaryBy(Age~agecat, data =data1, FUN=siterange)

# so only signed in users have ages and genders
summaryBy(Gender+Signed_In+Impressions+Clicks~agecat,
          data =data1)

# plot
install.packages("ggplot2")
library(ggplot2)
ggplot(data1, aes(x=Impressions, fill=agecat))
        +geom_histogram(binwidth=1)
ggplot(data1, aes(x=agecat, y=Impressions, fill=agecat))
        +geom_boxplot()

# create click thru rate
# we don't care about clicks if there are no impressions
# if there are clicks with no imps my assumptions about
# this data are wrong
data1$hasimps <-cut(data1$Impressions,c(-Inf,0,Inf))
summaryBy(Clicks~hasimps, data =data1, FUN=siterange)
ggplot(subset(data1, Impressions>0), aes(x=Clicks/Impressions,
```

```
        colour=agecat)) + geom_density()
ggplot(subset(data1, Clicks>0), aes(x=Clicks/Impressions,
        colour=agecat)) + geom_density()
ggplot(subset(data1, Clicks>0), aes(x=agecat, y=Clicks,
     fill=agecat)) + geom_boxplot()
ggplot(subset(data1, Clicks>0), aes(x=Clicks, colour=agecat))
     + geom_density()

# create categories
data1$scode[data1$Impressions==0] <- "NoImps"
data1$scode[data1$Impressions >0] <- "Imps"
data1$scode[data1$Clicks >0] <- "Clicks"

# Convert the column to a factor
data1$scode <- factor(data1$scode)
head(data1)

#look at levels
clen <- function(x){c(length(x))}
etable<-summaryBy(Impressions~scode+Gender+agecat,
                data = data1, FUN=clen)
```

Hint for doing the rest: don't read all the datasets into memory. Once you've perfected your code for one day, read the datasets in one at a time, process them, output any relevant metrics and variables, and store them in a dataframe; then remove the dataset before reading in the next one. This is to get you thinking about how to handle data sharded across multiple machines.

## On Coding

In a May 2013 op-ed piece, "How to be a Woman Programmer," Ellen Ullman describes quite well what it takes to be a programmer (setting aside for now the woman part):

"The first requirement for programming is a passion for the work, a deep need to probe the mysterious space between human thoughts and what a machine can understand; between human desires and how machines might satisfy them.

The second requirement is a high tolerance for failure. Programming is the art of algorithm design and the craft of debugging errant code. In the words of the great John Backus, inventor of the Fortran programming language: *You need the willingness to fail all the time. You have to generate many ideas and then you have to work very hard only to discover that they don't work. And you keep doing that over and over until you find one that does work."*

# The Data Science Process

Let's put it all together into what we define as the data science process. The more examples you see of people doing data science, the more you'll find that they fit into the general framework shown in Figure 2-2. As we go through the book, we'll revisit stages of this process and examples of it in different ways.

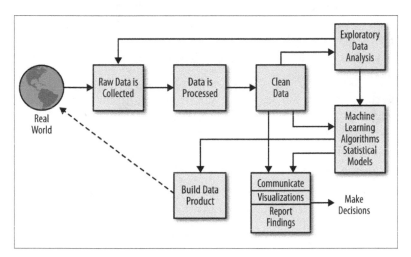

*Figure 2-2. The data science process*

First we have the Real World. Inside the Real World are lots of people busy at various activities. Some people are using Google+, others are competing in the Olympics; there are spammers sending spam, and there are people getting their blood drawn. Say we have data on one of these things.

Specifically, we'll start with raw data—logs, Olympics records, Enron employee emails, or recorded genetic material (note there are lots of aspects to these activities already lost even when we have that raw data). We want to process this to make it clean for analysis. So we build and use pipelines of data munging: joining, scraping, wrangling, or whatever you want to call it. To do this we use tools such as Python, shell scripts, R, or SQL, or all of the above.

Eventually we get the data down to a nice format, like something with columns:

name | event | year | gender | event time

 This is where you typically *start* in a standard statistics class, with a clean, orderly dataset. But it's not where you typically start in the real world.

Once we have this clean dataset, we should be doing some kind of EDA. In the course of doing EDA, we may realize that it isn't actually clean because of duplicates, missing values, absurd outliers, and data that wasn't actually logged or incorrectly logged. If that's the case, we may have to go back to collect more data, or spend more time cleaning the dataset.

Next, we design our model to use some algorithm like k-nearest neighbor (k-NN), linear regression, Naive Bayes, or something else. The model we choose depends on the type of problem we're trying to solve, of course, which could be a classification problem, a prediction problem, or a basic description problem.

We then can interpret, visualize, report, or communicate our results. This could take the form of reporting the results up to our boss or coworkers, or publishing a paper in a journal and going out and giving academic talks about it.

Alternatively, our goal may be to build or prototype a "data product"; e.g., a spam classifier, or a search ranking algorithm, or a recommendation system. Now the key here that makes data science special and distinct from statistics is that this data product then *gets incorporated back* into the real world, and users interact with that product, and that generates more data, which creates a feedback loop.

This is very different from predicting the weather, say, where your model doesn't influence the outcome at all. For example, you might predict it will rain next week, and unless you have some powers we don't know about, you're not going to *cause* it to rain. But if you instead build a recommendation system that generates evidence that "lots of people love this book," say, then you will know that you caused that feedback loop.

Take this loop into account in any analysis you do by adjusting for any biases your model caused. Your models are not just predicting the future, but *causing* it!

A data product that is productionized and that users interact with is at one extreme and the weather is at the other, but regardless of the type of data you work with and the "data product" that gets built on top of it—be it public policy determined by a statistical model, health insurance, or election polls that get widely reported and perhaps influence viewer opinions—you should consider the extent to which your model is influencing the very phenomenon that you are trying to observe and understand.

## A Data Scientist's Role in This Process

This model so far seems to suggest this will all magically happen without human intervention. By "human" here, we mean "data scientist." Someone has to make the decisions about what data to collect, and why. That person needs to be formulating questions and hypotheses and making a plan for how the problem will be attacked. And that someone is the data scientist or our beloved data science team.

Let's revise or at least add an overlay to make clear that the data scientist needs to be involved in this process throughout, meaning they are involved in the actual coding as well as in the higher-level process, as shown in Figure 2-3.

---

### Connection to the Scientific Method

We can think of the data science process as an extension or variation of the scientific method:

- Ask a question.
- Do background research.
- Construct a hypothesis.
- Test your hypothesis by doing an experiment.
- Analyze your data and draw a conclusion.
- Communicate your results.

In both the data science process and the scientific method, not every problem requires one to go through all the steps, but almost all prob-

---

lems can be solved with some combination of the stages. For example, if your end goal is a data visualization (which itself could be thought of as a data product), it's possible you might not do any machine learning or statistical modeling, but you'd want to get all the way to a clean dataset, do some exploratory analysis, and then create the visualization.

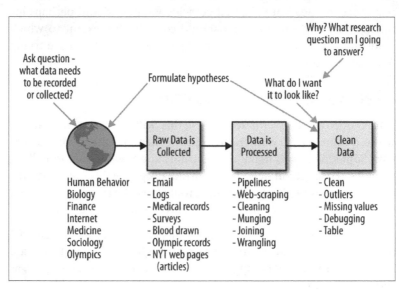

*Figure 2-3. The data scientist is involved in every part of this process*

# Thought Experiment: How Would You Simulate Chaos?

Most data problems start out with a certain amount of dirty data, ill-defined questions, and urgency. As data scientists we are, in a sense, attempting to create order from chaos. The class took a break from the lecture to discuss how they'd simulate chaos. Here are some ideas from the discussion:

- A Lorenzian water wheel, which is a Ferris wheel-type contraption with equally spaced buckets of water that rotate around in a circle. Now imagine water being dripped into the system at the very top. Each bucket has a leak, so some water escapes into whatever bucket is directly below the drip. Depending on the rate of the water coming in, this system exhibits a chaotic process that depends on molecular-level interactions of water molecules on the sides of the buckets. Read more about it in this associated Wikipedia article (*http://goo.gl/SjcJ64*).

- Many systems can exhibit inherent chaos. Philippe M. Binder and Roderick V. Jensen have written a paper entitled "Simulating chaotic behavior with finite-state machines" (*http://goo.gl/0LoaHw*), which is about digital computer simulations of chaos.

- An interdisciplinary program involving M.I.T., Harvard, and Tufts involved teaching a technique that was entitled "Simulating chaos to teach order" (*http://hvrd.me/1g3zzBz*). They simulated an emergency on the border between Chad and Sudan's troubled Darfur region, with students acting as members of Doctors Without Borders, International Medical Corps, and other humanitarian agencies.

- See also Joel Gascoigne's related essay, "Creating order from chaos in a startup" (*http://goo.gl/v43tZt*).

---

## Instructor Notes

1. Being a data scientist in an organization is often a chaotic experience, and it's the data scientist's job to try to create order from that chaos. So I wanted to simulate that chaotic experience for my students throughout the semester. But I also wanted them to know that things were going to be slightly chaotic for a pedagogical reason, and not due to my ineptitude!

2. I wanted to draw out different interpretations of the word "chaos" as a means to think about the importance of vocabulary, and the difficulties caused in communication when people either don't know what a word means, or have different ideas of what the word means. Data scientists might be communicating with domain experts who don't really understand what "logistic regression" means, say, but will pretend to know because they don't want to appear stupid, or because they think they ought to know, and therefore don't ask. But then the whole conversation is not really a successful communication if the two people talking don't really understand what they're talking about. Similarly, the data scientists ought to be asking questions to make sure they understand the terminology the domain expert is using (be it an astrophysicist, a social networking expert, or a climatologist).

---

There's nothing wrong with not knowing what a word means, but there is something wrong with not asking! You will likely find that asking clarifying questions about vocabulary gets you even more insight into the underlying data problem.

3. Simulation is a useful technique in data science. It can be useful practice to simulate fake datasets from a model to understand the generative process better, for example, and also to debug code.

# Case Study: RealDirect

Doug Perlson, the CEO of RealDirect, has a background in real estate law, startups, and online advertising. His goal with RealDirect is to use all the data he can access about real estate to improve the way people sell and buy houses.

Normally, people sell their homes about once every seven years, and they do so with the help of professional brokers and current data. But there's a problem both with the broker system and the data quality. RealDirect addresses both of them.

First, the brokers. They are typically "free agents" operating on their own—think of them as home sales consultants. This means that they guard their data aggressively, and the really good ones have lots of experience. But in the grand scheme of things, that really means they have only slightly more data than the inexperienced brokers.

RealDirect is addressing this problem by hiring a team of licensed real-estate agents who work together and pool their knowledge. To accomplish this, it built an interface for sellers, giving them useful data-driven tips on how to sell their house. It also uses interaction data to give real-time recommendations on what to do next.

The team of brokers also become data experts, learning to use information-collecting tools to keep tabs on new and relevant data or to access publicly available information. For example, you can now get data on co-op (a certain kind of apartment in NYC) sales, but that's a relatively recent change.

One problem with publicly available data is that it's old news—there's a three-month lag between a sale and when the data about that sale is available. RealDirect is working on real-time feeds on things like when people start searching for a home, what the initial offer is, the time between offer and close, and how people search for a home online.

Ultimately, good information helps both the buyer and the seller. At least if they're honest.

## How Does RealDirect Make Money?

First, it offers a subscription to sellers—about $395 a month—to access the selling tools. Second, it allows sellers to use RealDirect's agents at a reduced commission, typically 2% of the sale instead of the usual 2.5% or 3%. This is where the magic of data pooling comes in: it allows RealDirect to take a smaller commission because it's more optimized, and therefore gets more volume.

The site itself is best thought of as a platform for buyers and sellers to manage their sale or purchase process. There are statuses for each person on site: active, offer made, offer rejected, showing, in contract, etc. Based on your status, different actions are suggested by the software.

There are some challenges they have to deal with as well, of course. First off, there's a law in New York that says you can't show all the current housing listings unless those listings reside behind a registration wall, so RealDirect requires registration. On the one hand, this is an obstacle for buyers, but serious buyers are likely willing to do it. Moreover, places that don't require registration, like Zillow, aren't true competitors to RealDirect because they are merely showing listings without providing any additional service. Doug pointed out that you also need to register to use Pinterest, and it has tons of users in spite of this.

RealDirect comprises licensed brokers in various established realtor associations, but even so it has had its share of hate mail from realtors who don't appreciate its approach to cutting commission costs. In this sense, RealDirect is breaking directly into a guild. On the other hand, if a realtor refused to show houses because they are being sold on RealDirect, the potential buyers would see those listings elsewhere and complain. So the traditional brokers have little choice but to deal with RealDirect even if they don't like it. In other words, the listings themselves are sufficiently transparent so that the traditional brokers can't get away with keeping their buyers away from these houses.

Doug talked about key issues that a buyer might care about—nearby parks, subway, and schools, as well as the comparison of prices per square foot of apartments sold in the same building or block. This is the kind of data they want to increasingly cover as part of the service of RealDirect.

## Exercise: RealDirect Data Strategy

You have been hired as chief data scientist at *realdirect.com*, and report directly to the CEO. The company (hypothetically) does not yet have its data plan in place. It's looking to you to come up with a data strategy. Here are a couple ways you could begin to approach this problem:

1. Explore its existing website, thinking about how buyers and sellers would navigate through it, and how the website is structured/organized. Try to understand the existing business model, and think about how analysis of RealDirect user-behavior data could be used to inform decision-making and product development. Come up with a list of research questions you think could be answered by data:

   - What data would you advise the engineers log and what would your ideal datasets look like?

   - How would data be used for reporting and monitoring product usage?

   - How would data be built back into the product/website?

2. Because there is no data yet for you to analyze (typical in a start-up when its still building its product), you should get some auxiliary data to help gain intuition about this market. For example, go to *https://github.com/oreillymedia/doing_data_science*. Click on Rolling Sales Update (after the fifth paragraph).

   You can use any or all of the datasets here—start with Manhattan August, 2012–August 2013.

   - First challenge: load in and clean up the data. Next, conduct exploratory data analysis in order to find out where there are outliers or missing values, decide how you will treat them, make sure the dates are formatted correctly, make sure values you think are numerical are being treated as such, etc.

   - Once the data is in good shape, conduct exploratory data analysis to visualize and make comparisons (i) across neighborhoods, and (ii) across time. If you have time, start looking for meaningful patterns in this dataset.

3. Summarize your findings in a brief report aimed at the CEO.

4. Being the "data scientist" often involves speaking to people who aren't also data scientists, so it would be ideal to have a set of communication strategies for getting to the information you need about the data. Can you think of any other people you should talk to?

5. Most of you are not "domain experts" in real estate or online businesses.

   - Does stepping out of your comfort zone and figuring out how you would go about "collecting data" in a different setting give you insight into how you do it in your own field?

   - Sometimes "domain experts" have their own set of vocabulary. Did Doug use vocabulary specific to his domain that you didn't understand ("comps," "open houses," "CPC")? Sometimes if you don't understand vocabulary that an expert is using, it can prevent you from understanding the problem. It's good to get in the habit of asking questions because eventually you will get to something you do understand. This involves persistence and is a habit to cultivate.

6. Doug mentioned the company didn't necessarily have a data strategy. There is no industry standard for creating one. As you work through this assignment, think about whether there is a set of best practices you would recommend with respect to developing a data strategy for an online business, or in your own domain.

## Sample R code

Here's some sample R code that takes the Brooklyn housing data in the preceding exercise, and cleans and explores it a bit. (The exercise asks you to do this for Manhattan.)

```
# Author: Benjamin Reddy

library(plyr)

require(gdata)
bk <- read.xls("rollingsales_brooklyn.xls",pattern="BOROUGH")
head(bk)
summary(bk)

bk$SALE.PRICE.N <- as.numeric(gsub("[^[:digit:]]","",
```

```
                              bk$SALE.PRICE))
count(is.na(bk$SALE.PRICE.N))

names(bk) <- tolower(names(bk))

## clean/format the data with regular expressions
bk$gross.sqft <- as.numeric(gsub("[^[:digit:]]","",
                              bk$gross.square.feet))
bk$land.sqft <- as.numeric(gsub("[^[:digit:]]","",
                              bk$land.square.feet))

bk$sale.date <- as.Date(bk$sale.date)
bk$year.built <- as.numeric(as.character(bk$year.built))

## do a bit of exploration to make sure there's not anything
## weird going on with sale prices
attach(bk)

hist(sale.price.n)
hist(sale.price.n[sale.price.n>0])
hist(gross.sqft[sale.price.n==0])

detach(bk)

## keep only the actual sales
bk.sale <- bk[bk$sale.price.n!=0,]

plot(bk.sale$gross.sqft,bk.sale$sale.price.n)
plot(log(bk.sale$gross.sqft),log(bk.sale$sale.price.n))

## for now, let's look at 1-, 2-, and 3-family homes
bk.homes <- bk.sale[which(grepl("FAMILY",
            bk.sale$building.class.category)),]
plot(log(bk.homes$gross.sqft),log(bk.homes$sale.price.n))

bk.homes[which(bk.homes$sale.price.n<100000),]
    [order(bk.homes[which(bk.homes$sale.price.n<100000),]
            $sale.price.n),]

## remove outliers that seem like they weren't actual sales
bk.homes$outliers <- (log(bk.homes$sale.price.n) <=5) + 0
bk.homes <- bk.homes[which(bk.homes$outliers==0),]

plot(log(bk.homes$gross.sqft),log(bk.homes$sale.price.n))
```

# CHAPTER 3

# Algorithms

In the previous chapter we discussed in general how models are used in data science. In this chapter, we're going to be diving into algorithms.

An algorithm is a procedure or set of steps or rules to accomplish a task. Algorithms are one of the fundamental concepts in, or building blocks of, computer science: the basis of the design of elegant and efficient code, data preparation and processing, and software engineering.

Some of the basic types of tasks that algorithms can solve are sorting, searching, and graph-based computational problems. Although a given task such as sorting a list of objects could be handled by multiple possible algorithms, there is some notion of "best" as measured by efficiency and computational time, which matters especially when you're dealing with massive amounts of data and building consumer-facing products.

Efficient algorithms that work sequentially or in parallel are the basis of pipelines to process and prepare data. With respect to data science, there are at least three classes of algorithms one should be aware of:

1. Data munging, preparation, and processing algorithms, such as sorting, MapReduce, or Pregel.

   We would characterize these types of algorithms as data engineering, and while we devote a chapter to this, it's not the emphasis of this book. This is not to say that you won't be doing data wrangling and munging—just that we don't emphasize the algorithmic aspect of it.

2. Optimization algorithms for parameter estimation, including Stochastic Gradient Descent, Newton's Method, and Least Squares. We mention these types of algorithms throughout the book, and they underlie many R functions.

3. Machine learning algorithms are a large part of this book, and we discuss these more next.

# Machine Learning Algorithms

Machine learning algorithms are largely used to predict, classify, or cluster.

Wait! Back in the previous chapter, didn't we already say modeling could be used to predict or classify? Yes. Here's where some lines have been drawn that can make things a bit confusing, and it's worth understanding who drew those lines.

Statistical *modeling* came out of statistics departments, and machine learning *algorithms* came out of computer science departments. Certain methods and techniques are considered to be part of both, and you'll see that we often use the words somewhat interchangeably.

You'll find some of the methods in this book, such as linear regression, in machine learning books as well as intro to statistics books. It's not necessarily useful to argue over who the rightful owner is of these methods, but it's worth pointing out here that it can get a little vague or ambiguous about what the actual difference is.

In general, machine learning algorithms that are the basis of artificial intelligence (AI) such as image recognition, speech recognition, recommendation systems, ranking and personalization of content— often the basis of data products—are not usually part of a core statistics curriculum or department. They aren't generally designed to infer the underlying *generative process* (e.g., to model something), but rather to predict or classify with the most accuracy.

These differences in methods reflect in cultural differences in the approaches of machine learners and statisticians that Rachel observed at Google, and at industry conferences. Of course, data scientists can and should use both approaches.

There are some broad generalizations to consider:

*Interpreting parameters*

Statisticians think of the parameters in their linear regression models as having real-world interpretations, and typically want to be able to find meaning in behavior or describe the real-world phenomenon corresponding to those parameters. Whereas a software engineer or computer scientist might be wanting to build their linear regression algorithm into production-level code, and the predictive model is what is known as a *black box algorithm*, they don't generally focus on the interpretation of the parameters. If they do, it is with the goal of handtuning them in order to optimize *predictive power*.

*Confidence intervals*

Statisticians provide confidence intervals and posterior distributions for parameters and estimators, and are interested in capturing the variability or uncertainty of the parameters. Many machine learning algorithms, such as k-means or k-nearest neighbors (which we cover a bit later in this chapter), don't have a notion of confidence intervals or uncertainty.

*The role of explicit assumptions*

Statistical models make explicit assumptions about data-generating processes and distributions, and you use the data to estimate parameters. Nonparametric solutions, like we'll see later in this chapter, don't make any assumptions about probability distributions, or they are implicit.

We say the following lovingly and with respect: statisticians have chosen to spend their lives investigating uncertainty, and they're never 100% confident about anything. Software engineers like to build things. They want to build models that predict the best they can, but there are no concerns about uncertainty—just build it! At companies like Facebook or Google, the philosophy is to build and iterate often. If something breaks, it can be fixed. A data scientist who somehow manages to find a balance between the statistical and computer science approaches, and to find value in both these ways of being, can thrive. Data scientists are the multicultural statistician-computer scientist hybrid, so we're not tied to any one way of thinking over another; they both have value. We'll sum up our take on this with guest speaker Josh Wills' (Chapter 13) well-tweeted quote:

Data scientist (noun): Person who is better at statistics than any software engineer and better at software engineering than any statistician.

— Josh Wills

# Three Basic Algorithms

Many business or real-world problems that can be solved with data can be thought of as *classification* and *prediction* problems when we express them mathematically. Happily, a whole host of models and algorithms can be used to classify and predict.

Your real challenge as a data scientist, once you've become familiar with how to implement them, is understanding which ones to use depending on the context of the problem and the underlying assumptions. This partially comes with experience—you start seeing enough problems that you start thinking, "Ah, this is a classification problem with a binary outcome" or, "This is a classification problem, but oddly I don't even have any labels" and you know what to do. (In the first case, you could use logistic regression or Naive Bayes, and in the second you could start with k-means—more on all these shortly!)

Initially, though, when you hear about these methods in isolation, it takes some effort on your part as a student or learner to think, "In the real world, how do I know that this algorithm is the solution to the problem I'm trying to solve?"

It's a real mistake to be the type of person walking around with a hammer looking for a nail to bang: "I know linear regression, so I'm going to try to solve every problem I encounter with it." Don't do that. Instead, try to understand the context of the problem, and the attributes it has *as a problem*. Think of those in mathematical terms, and then think about the algorithms you know and how they map to this type of problem.

If you're not sure, it's good to talk it through with someone who does. So ask a coworker, head to a meetup group, or start one in your area! Also, maintain the attitude that it's *not obvious* what to do and that's what makes it a problem, and so you're going to approach it circumspectly and methodically. You don't have to be the know-it-all in the room who says, "Well, *obviously* we should use linear regression with a penalty function for regularization," even if that seems to you the right approach.

We're saying all this because one of the unfortunate aspects of textbooks is they often give you a bunch of techniques and then problems that tell you *which* method to use that solves the problem (e.g., use linear regression to predict height from weight). Yes, implementing and understanding linear regression the first few times is not obvious, so you need practice with that, but it needs to be addressed that the real challenge once you have mastery over technique is *knowing when to use linear regression in the first place*.

We're not going to give a comprehensive overview of *all* possible machine learning algorithms, because that would make this a machine learning book, and there are already plenty of those.

Having said that, in this chapter we'll introduce three basic algorithms now and introduce others throughout the book in context. By the end of the book, you should feel more confident about your ability to learn new algorithms so that you can pick them up along the way as problems require them.

We'll also do our best to demonstrate the thought processes of data scientists who had to figure out which algorithm to use in context and why, but it's also upon you as a student and learner to *force yourself* to think about what the attributes of the problem were that made a given algorithm the right algorithm to use.

With that said, we still need to give you some basic tools to use, so we'll start with linear regression, k-nearest neighbors (k-NN), and k-means. In addition to what was just said about trying to understand the attributes of problems that could use these as solutions, look at these three algorithms from the perspective of: what patterns can we as humans see in the data with our eyes that we'd like to be able to automate with a machine, especially taking into account that as the data gets more complex, we can't see these patterns?

## Linear Regression

One of the most common statistical methods is linear regression. At its most basic, it's used when you want to express the mathematical relationship between two variables or attributes. When you use it, you are making the assumption that there is a *linear* relationship between an outcome variable (sometimes also called the response variable, dependent variable, or label) and a predictor (sometimes also called an independent variable, explanatory variable, or feature); or between one variable and several other variables, in which case you're *modeling* the relationship as having a linear structure.

## WTF. So Is It an Algorithm or a Model?

While we tried to make a distinction between the two earlier, we admit the colloquial use of the words "model" and "algorithm" gets confusing because the two words seem to be used interchangeably when their actual definitions are not the same thing at all. In the purest sense, an algorithm is a set of rules or steps to follow to accomplish some task, and a model is an attempt to describe or capture the world. These two seem obviously different, so it seems the distinction should should be obvious. Unfortunately, it isn't. For example, regression can be described as a statistical model as well as a machine learning algorithm. You'll waste your time trying to get people to discuss this with any precision.

In some ways this is a historical artifact of statistics and computer science communities developing methods and techniques in parallel and using different words for the same methods. The consequence of this is that the distinction between machine learning and statistical modeling is muddy. Some methods (for example, k-means, discussed in the next section) we might call an *algorithm* because it's a series of computational steps used to cluster or classify objects—on the other hand, k-means can be reinterpreted as a special case of a Gaussian mixture *model*. The net result is that colloquially, people use the terms algorithm and model interchangeably when it comes to a lot of these methods, so try not to let it worry you. (Though it bothers us, too.)

Assuming that there is a *linear* relationship between an outcome variable and a predictor is a big assumption, but it's also the simplest one you *can* make—linear functions are more basic than nonlinear ones in a mathematical sense—so in that sense it's a good starting point.

In some cases, it makes sense that changes in one variable correlate linearly with changes in another variable. For example, it makes sense that the more umbrellas you sell, the more money you make. In those cases you can feel good about the linearity assumption. Other times, it's harder to justify the assumption of linearity except locally: in the spirit of calculus, everything can be approximated by line segments as long as functions are continuous.

Let's back up. Why would you even want to build a linear model in the first place? You might want to use this relationship to *predict* future outcomes, or you might want to understand or *describe* the relationship to get a grasp on the situation. Let's say you're studying the relationship between a company's sales and how much that company spends on advertising, or the number of friends someone has on a social networking site and the time that person spends on that site daily. These are all numerical outcomes, which mean linear regression would be a wise choice, at least for a first pass at your problem.

One entry point for thinking about linear regression is to think about deterministic lines first. We learned back in grade school that we could describe a line with a slope and an intercept, $y = f(x) = \beta_0 + \beta_1 * x$. But the setting there was always deterministic.

Even for the most mathematically sophisticated among us, if you haven't done it before, it's a new mindset to start thinking about stochastic functions. We still have the same components: points listed out explicitly in a table (or as tuples), and functions represented in equation form or plotted on a graph. So let's build up to linear regression starting from a deterministic function.

**Example 1. Overly simplistic example to start**. Suppose you run a social networking site that charges a monthly subscription fee of $25, and that this is your only source of revenue. Each month you collect data and count your number of users and total revenue. You've done this daily over the course of two years, recording it all in a spreadsheet. You could express this data as a series of points. Here are the first four:

$$S = \{(x, y) = (1, 25), (10, 250), (100, 2500), (200, 5000)\}$$

If you showed this to someone else who didn't even know how much you charged or anything about your business model (what kind of friend wasn't paying attention to your business model?!), they might notice that there's a clear relationship enjoyed by all of these points, namely $y = 25x$. They likely could do this in their head, in which case they figured out that:

- There's a linear pattern.
- The coefficient relating $x$ and $y$ is 25.
- It seems deterministic.

You can even plot it as in Figure 3-1 to verify they were right (even though you knew they were because you made the business model in the first place). It's a line!

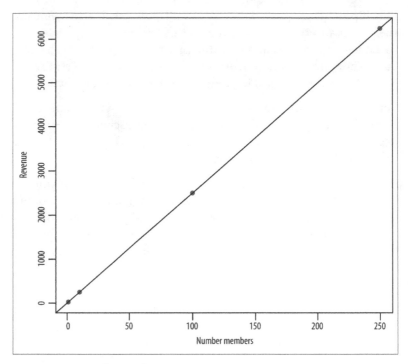

*Figure 3-1. An obvious linear pattern*

**Example 2. Looking at data at the user level.** Say you have a dataset *keyed* by user (meaning each row contains data for a single user), and the columns represent user behavior on a social networking site over a period of a week. Let's say you feel comfortable that the data is clean at this stage and that you have on the order of hundreds of thousands of users. The names of the columns are total_num_friends, total_new_friends_this_week, num_visits, time_spent, number_apps_downloaded, number_ads_shown, gender, age, and so on. During the course of your exploratory data analysis, you've randomly sampled 100 users to keep it simple, and you plot pairs of these variables, for example, $x$ = total_new_friends and $y$ = time_spent (in seconds). The business context might be that eventually you want to be able to promise advertisers who bid for space on your website in advance a certain number of users, so you want to be able to forecast number of users several days or weeks in advance. But for now, you are simply trying to build intuition and understand your dataset.

You eyeball the first few rows and see:

| | |
|---|---|
| 7 | 276 |
| 3 | 43 |
| 4 | 82 |
| 6 | 136 |
| 10 | 417 |
| 9 | 269 |

Now, your brain can't figure out what's going on by just looking at them (and your friend's brain probably can't, either). They're in no obvious particular order, and there are a lot of them. So you try to plot it as in Figure 3-2.

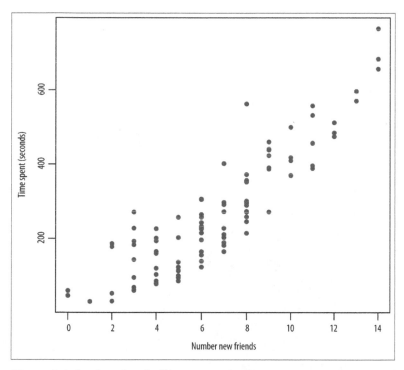

Figure 3-2. Looking kind of linear

It looks like there's *kind of* a linear relationship here, and it makes sense; the more new friends you have, the more time you might spend on the site. But how can you figure out how to describe that relationship? Let's also point out that there is no perfectly *deterministic* relationship between number of new friends and time spent on the site, but it makes sense that there is an *association* between these two variables.

### Start by writing something down

There are two things you want to capture in the model. The first is the *trend* and the second is the *variation*. We'll start first with the trend.

First, let's start by assuming there actually *is* a relationship and that it's linear. It's the best you can do at this point.

There are many lines that look more or less like they might work, as shown in Figure 3-3.

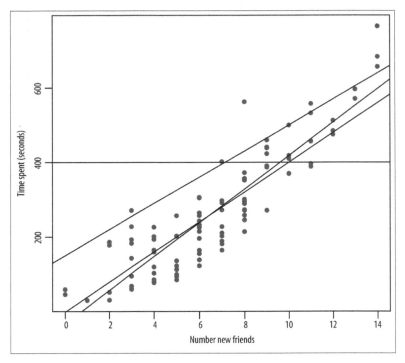

*Figure 3-3. Which line is the best fit?*

So how do you pick which one?

Because you're assuming a linear relationship, start your model by assuming the functional form to be:

$$y = \beta_0 + \beta_1 x$$

Now your job is to find the best choices for $\beta_0$ and $\beta_1$ using the observed data to estimate them: $(x_1, y_1), (x_2, y_2), \ldots (x_n, y_n)$.

Writing this with matrix notation results in this:

$$y = x \cdot \beta$$

There you go: you've written down your model. Now the rest is *fitting* the model.

### Fitting the model

So, how do you calculate $\beta$? The intuition behind linear regression is that you want to find the line that minimizes the distance between all the points and the line.

Many lines look approximately correct, but your goal is to find the optimal one. Optimal could mean different things, but let's start with optimal to mean the line that, on average, is closest to all the points. But what does *closest* mean here?

Look at Figure 3-4. Linear regression seeks to find the line that minimizes the sum of the squares of the vertical distances between the approximated or predicted $\hat{y}_i$s and the observed $y_i$s. You do this because you want to minimize your prediction errors. This method is called *least squares* estimation.

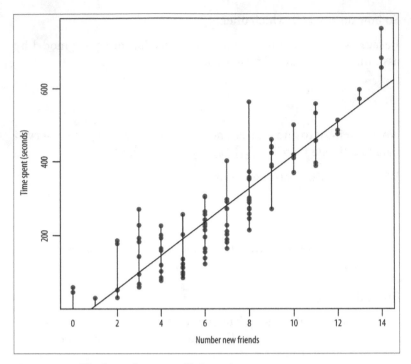

*Figure 3-4. The line closest to all the points*

To find this line, you'll define the "residual sum of squares" (RSS), denoted $RSS(\beta)$, to be:

$$RSS(\beta) = \Sigma_i \left( y_i - \beta x_i \right)^2$$

where $i$ ranges over the various data points. It is the sum of all the squared vertical distances between the observed points and any given line. Note this is a function of $\beta$ and you want to optimize with respect to $\beta$ to find the optimal line.

To minimize $RSS(\beta) = (y - \beta x)^t (y - \beta x)$, differentiate it with respect to $\beta$ and set it equal to zero, then solve for $\beta$. This results in:

$$\hat{\beta} = \left( x^t x \right)^{-1} x^t y$$

Here the little "hat" symbol on top of the $\beta$ is there to indicate that it's the *estimator* for $\beta$. You don't know the true value of $\beta$; all you have is the observed data, which you plug into the estimator to get an estimate.

To actually fit this, to get the $\beta$s, all you need is one line of R code where you've got a column of y's and a (single) column of x's:

```
model <- lm(y ~ x)
```

So for the example where the first few rows of the data were:

| x | y |
|---|---|
| 7 | 276 |
| 3 | 43 |
| 4 | 82 |
| 6 | 136 |
| 10 | 417 |
| 9 | 269 |

The R code for this would be:

```
> model <- lm (y~x)
> model

Call:
lm(formula = y ~ x)

Coefficients:
(Intercept)            x
     -32.08        45.92

> coefs <- coef(model)
> plot(x, y, pch=20,col="red", xlab="Number new friends",
  ylab="Time spent (seconds)")
> abline(coefs[1],coefs[2])
```

And the estimated line is $\hat{y} = -32.08 + 45.92x$, which you're welcome to round to $\hat{y} = -32 + 46x$, and the corresponding plot looks like the lefthand side of Figure 3-5.

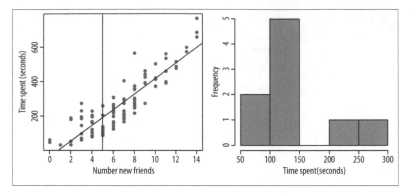

*Figure 3-5. On the left is the fitted line. We can see that for any fixed value, say 5, the values for y vary. For people with 5 new friends, we display their time spent in the plot on the right.*

But it's up to you, the data scientist, whether you think you'd actually want to use this linear model to describe the relationship or predict new outcomes. If a new x-value of 5 came in, meaning the user had five new friends, how confident are you in the output value of –32.08 + 45.92*5 = 195.7 seconds?

In order to get at this question of confidence, you need to extend your model. You know there's variation among time spent on the site by people with five new friends, meaning you certainly wouldn't make the claim that everyone with five new friends is guaranteed to spend 195.7 seconds on the site. So while you've so far modeled the *trend*, you haven't yet modeled the *variation*.

### Extending beyond least squares

Now that you have a *simple linear regression model* down (one output, one predictor) using least squares estimation to estimate your $\beta$s, you can build upon that model in three primary ways, described in the upcoming sections:

1. Adding in modeling assumptions about the errors

2. Adding in more predictors

3. Transforming the predictors

**Adding in modeling assumptions about the errors.** If you use your model to predict $y$ for a given value of $x$, your prediction is deterministic and doesn't capture the variablility in the observed data. See on the righthand side of Figure 3-5 that for a fixed value of $x = 5$, there is variability among the time spent on the site. You want to capture this variability in your model, so you extend your model to:

$$y = \beta_0 + \beta_1 x + \epsilon$$

where the new term $\epsilon$ is referred to as *noise*, which is the stuff that you haven't accounted for by the relationships you've figured out so far. It's also called the *error term*—$\epsilon$ represents the *actual error*, the difference between the observations and the *true* regression line, which you'll never know and can only estimate with your $\hat{\beta}$s.

One often makes the modeling assumption that the noise is normally distributed, which is denoted:

$$\epsilon \sim N\!\left(0, \sigma^2\right)$$

Note this is sometimes not a reasonable assumption. If you are dealing with a known fat-tailed distribution, and if your linear model is picking up only a small part of the value of the variable y, then the error terms are likely also fat-tailed. This is the most common situation in financial modeling.

That's not to say we don't use linear regression in finance, though. We just don't attach the "noise is normal" assumption to it.

With the preceding assumption on the distribution of noise, this model is saying that, for any given value of $x$, the conditional distribution of $y$ given $x$ is $p(y|x) \sim N\!\left(\beta_0 + \beta_1 x, \sigma^2\right)$.

So, for example, among the set of people who had five new friends this week, the amount of the time they spent on the website had a *normal distribution* with a mean of $\beta_0 + \beta_1 * 5$ and a variance of $\sigma^2$, and you're going to estimate your parameters $\beta_0, \beta_1, \sigma$ from the data.

How do you fit this model? How do you get the parameters $\beta_0, \beta_1, \sigma$ from the data?

Turns out that no matter how the εs are distributed, the least squares estimates that you already derived are the optimal estimators for βs because they have the property of being unbiased and of being the minimum variance estimators. If you want to know more about these properties and see a proof for this, we refer you to any good book on statistical inference (for example, *Statistical Inference* by Casella and Berger).

So what can you do with your observed data to estimate the variance of the errors? Now that you have the estimated line, you can see how far away the observed data points are from the line itself, and you can treat these differences, also known as *observed errors* or *residuals* ,as observations themselves, or estimates of the actual errors, the εs. Define $e_i = y_i - \hat{y}_i = y_i - \left( \hat{\beta}_0 + \hat{\beta}_1 x_i \right)$ for $i = 1, \ldots, n$.

Then you estimate the variance ($\sigma^2$) of ε, as:

$$\frac{\sum_i e_i^2}{n-2}$$

Why are we dividing by $n-2$? A natural question. Dividing by $n-2$, rather than just $n$, produces an *unbiased estimator*. The 2 corresponds to the number of model parameters. Here again, Casella and Berger's book is an excellent resource for more background information.

This is called the *mean squared error* and captures how much the predicted value varies from the observed. *Mean squared error* is a useful quantity for any prediction problem. In regression in particular, it's *also* an estimator for your variance, but it can't always be used or interpreted that way. It appears in the evaluation metrics in the following section.

### Evaluation metrics

We asked earlier how confident you would be in these estimates and in your model. You have a couple values in the output of the R function that help you get at the issue of how confident you can be in the estimates: p-values and R-squared. Going back to our model in R, if we type in `summary(model)`, which is the name we gave to this model, the output would be:

```
summary (model)
Call:
lm(formula = y ~ x)

Residuals:
    Min      1Q  Median      3Q     Max
-121.17  -52.63   -9.72   41.54  356.27

Coefficients:
            Estimate Std. Error t value Pr(>|t|)
(Intercept)  -32.083     16.623   -1.93   0.0565 .
x             45.918      2.141   21.45   <2e-16 ***
Signif. codes:  0 '***' 0.001 '**' 0.01 '*' 0.05 '.' 0.1 ' ' 1

Residual standard error: 77.47 on 98 degrees of freedom
Multiple R-squared: 0.8244,    Adjusted R-squared: 0.8226
F-statistic:    460 on 1 and 98 DF,  p-value: < 2.2e-16
```

*R-squared*

$R^2 = 1 - \frac{\Sigma_i\left(y_i - \hat{y_i}\right)^2}{\Sigma_i\left(y_i - \bar{y}\right)^2}$. This can be interpreted as the proportion of variance explained by our model. Note that mean squared error is in there getting divided by total error, which is the proportion of variance *unexplained* by our model and we calculate 1 minus that.

*p-values*

Looking at the output, the estimated $\beta$s are in the column marked Estimate. To see the p-values, look at $Pr(>|t|)$. We can interpret the values in this column as follows: We are making a null hypothesis that the $\beta$s are zero. For any given $\beta$, the p-value captures the probability of observing the data that we observed, and obtaining the test-statistic that we obtained *under the null hypothesis*. This means that if we have a low p-value, it is highly unlikely to observe such a test-statistic under the null hypothesis, and the coefficient is highly likely to be nonzero and therefore significant.

*Cross-validation*

Another approach to evaluating the model is as follows. Divide our data up into a training set and a test set: 80% in the training and 20% in the test. Fit the model on the training set, then look at the *mean squared error* on the test set and compare it to that on the training set. Make this comparison across sample size as well. If the mean squared errors are approximately the same, then our model generalizes well and we're not in danger of overfitting. See Figure 3-6 to see what this might look like. This approach is highly recommended.

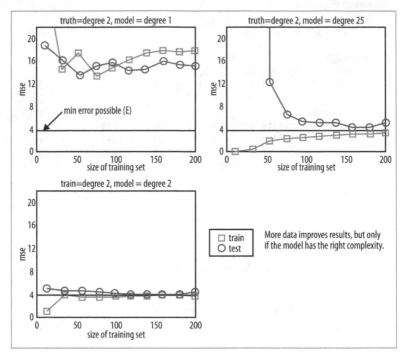

*Figure 3-6. Comparing mean squared error in training and test set, taken from a slide of Professor Nando de Freitas; here, the ground truth is known because it came from a dataset with data simulated from a known distribution*

### Other models for error terms

The mean squared error is an example of what is called a *loss function*. This is the standard one to use in linear regression because it gives us a pretty nice measure of closeness of fit. It has the additional desirable property that by assuming that εs are normally distributed, we can rely on the maximum likelihood principle. There are other loss functions such as one that relies on absolute value rather than squaring. It's also possible to build custom loss functions specific to your particular problem or context, but for now, you're safe with using mean square error.

**Adding other predictors.** What we just looked at was simple linear regression—one outcome or dependent variable and one predictor. But we can extend this model by building in other predictors, which is called *multiple linear regression*:

$$y = \beta_0 + \beta_1 x_1 + \beta_2 x_2 + \beta_3 x_3 + \epsilon.$$

All the math that we did before holds because we had expressed it in matrix notation, so it was already generalized to give the appropriate estimators for the $\beta$. In the example we gave of predicting time spent on the website, the other predictors could be the user's age and gender, for example. We'll explore feature selection more in Chapter 7, which means figuring out which additional predictors you'd want to put in your model. The R code will just be:

```
model <- lm(y ~ x_1 + x_2 + x_3)
```

Or to add in interactions between variables:

```
model <- lm(y ~ x_1 + x_2*x_3)
```

One key here is to make scatterplots of $y$ against each of the predictors as well as between the predictors, and histograms of $y|x$ for various values of each of the predictors to help build intuition. As with simple linear regression, you can use the same methods to evaluate your model as described earlier: looking at $R^2$, p-values, and using training and testing sets.

**Transformations.** Going back to one $x$ predicting one $y$, why did we assume a linear relationship? Instead, maybe, a better model would be a polynomial relationship like this:

$$y = \beta_0 + \beta_1 x + \beta_2 x^2 + \beta_3 x^3$$

Wait, but isn't this *linear* regression? Last time we checked, polynomials weren't linear. To think of it as *linear*, you transform or create new variables—for example, $z = x^2$—and build a regression model based on $z$. Other common transformations are to take the log or to pick a threshold and turn it into a binary predictor instead.

If you look at the plot of time spent versus number friends, the shape looks a little bit curvy. You could potentially explore this further by building up a model and checking to see whether this yields an improvement.

What you're facing here, though, is one of the biggest challenges for a modeler: you never know the truth. It's possible that the true model is quadratic, but you're assuming linearity or vice versa. You do your best to evaluate the model as discussed earlier, but you'll never *really* know if you're right. More and more data can sometimes help in this regard as well.

## Review

Let's review the assumptions we made when we built and fit our model:

- Linearity
- Error terms normally distributed with mean 0
- Error terms independent of each other
- Error terms have constant variance across values of $x$
- The predictors we're using are the *right* predictors

When and why do we perform linear regression? Mostly for two reasons:

- If we want to predict one variable knowing others
- If we want to explain or understand the relationship between two or more things

## Exercise

To help understand and explore new concepts, you can simulate fake datasets in R. The advantage of this is that you "play God" because you actually know the underlying truth, and you get to see how good your model is at recovering the truth.

Once you've better understood what's going on with your fake dataset, you can then transfer your understanding to a real one. We'll show you how to simulate a fake dataset here, then we'll give you some ideas for how to explore it further:

```
# Simulating fake data
x_1 <- rnorm(1000,5,7) # from a normal distribution simulate
                       # 1000 values with a mean of 5 and
                       #  standard deviation of 7
hist(x_1, col="grey") # plot p(x)
true_error <- rnorm(1000,0,2)
true_beta_0 <- 1.1
true_beta_1 <- -8.2

y <- true_beta_0 + true_beta_1*x_1 + true_error
```

```
hist(y) # plot p(y)
plot(x_1,y, pch=20,col="red") # plot p(x,y)
```

1. Build a regression model and see that it recovers the true values of the $\beta$s.

2. Simulate another fake variable $x_2$ that has a Gamma distribution with parameters you pick. Now make the truth be that $y$ is a linear combination of both $x_1$ and $x_2$. Fit a model that only depends on $x_1$. Fit a model that only depends on $x_2$. Fit a model that uses both. Vary the sample size and make a plot of mean square error of the training set and of the test set versus sample size.

3. Create a new variable, $z$, that is equal to $x_1^2$. Include this as one of the predictors in your model. See what happens when you fit a model that depends on $x_1$ only and then also on $z$. Vary the sample size and make a plot of mean square error of the training set and of the test set versus sample size.

4. Play around more by (a) changing parameter values (the true $\beta$s), (b) changing the distribution of the true error, and (c) including more predictors in the model with other kinds of probability distributions. (`rnorm()` means randomly generate values from a normal distribution. `rbinom()` does the same for binomial. So look up these functions online and try to find more.)

5. Create scatterplots of all pairs of variables and histograms of single variables.

## k-Nearest Neighbors (k-NN)

K-NN is an algorithm that can be used when you have a bunch of objects that have been classified or labeled in some way, and other similar objects that haven't gotten classified or labeled yet, and you want a way to automatically label them.

The objects could be data scientists who have been classified as "sexy" or "not sexy"; or people who have been labeled as "high credit" or "low credit"; or restaurants that have been labeled "five star," "four star," "three star," "two star," "one star," or if they really suck, "zero stars." More seriously, it could be patients who have been classified as "high cancer risk" or "low cancer risk."

Take a second and think whether or not linear regression would work to solve problems of this type.

OK, so the answer is: it depends. When you use linear regression, the output is a continuous variable. Here the output of your algorithm is going to be a categorical label, so linear regression wouldn't solve the problem as it's described.

However, it's not impossible to solve it with linear regression plus the concept of a "threshold." For example, if you're trying to predict people's credit scores from their ages and incomes, and then picked a threshold such as 700 such that if your prediction for a given person whose age and income you observed was above 700, you'd label their predicted credit as "high," or toss them into a bin labeled "high." Otherwise, you'd throw them into the bin labeled "low." With more thresholds, you could also have more fine-grained categories like "very low," "low," "medium," "high," and "very high."

In order to do it this way, with linear regression you'd have to establish the bins as ranges of a continuous outcome. But not everything is on a continuous scale like a credit score. For example, what if your labels are "likely Democrat," "likely Republican," and "likely independent"? What do you do now?

The intuition behind k-NN is to consider the *most similar* other items defined in terms of their attributes, look at their labels, and give the unassigned item the majority vote. If there's a tie, you randomly select among the labels that have tied for first.

So, for example, if you had a bunch of movies that were labeled "thumbs up" or "thumbs down," and you had a movie called "Data Gone Wild" that hadn't been rated yet—you could look at its attributes: length of movie, genre, number of sex scenes, number of Oscar-winning actors in it, and budget. You could then find other movies with similar attributes, look at *their* ratings, and then give "Data Gone Wild" a rating without ever having to watch it.

To automate it, two decisions must be made: first, how do you define *similarity* or closeness? Once you define it, for a given unrated item, you can say how similar *all* the labeled items are to it, and you can take the *most similar* items and call them *neighbors*, who each have a "vote."

This brings you to the second decision: how many neighbors should you look at or "let vote"? This value is $k$, which ultimately you'll choose as the data scientist, and we'll tell you how.

Make sense? Let's try it out with a more realistic example.

## Example with credit scores

Say you have the age, income, and a credit category of high or low for a bunch of people and you want to use the age and income to predict the credit label of "high" or "low" for a new person.

For example, here are the first few rows of a dataset, with income represented in thousands:

```
age income credit
 69      3    low
 66     57    low
 49     79    low
 49     17    low
 58     26   high
 44     71   high
```

You can plot people as points on the plane and label people with an empty circle if they have low credit ratings, as shown in Figure 3-7.

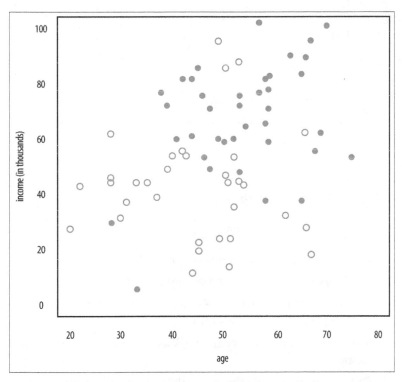

*Figure 3-7. Credit rating as a function of age and income*

What if a new guy comes in who is 57 years old and who makes $37,000? What's his likely credit rating label? Look at Figure 3-8. Based on the other people near him, what credit score label do you think he should be given? Let's use k-NN to do it automatically.

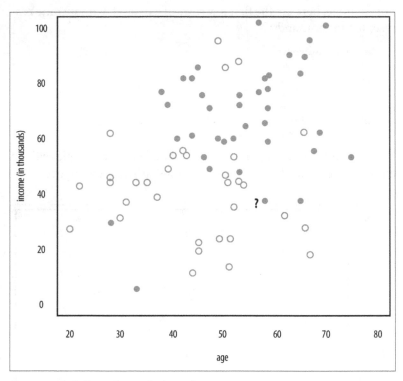

*Figure 3-8. What about that guy?*

Here's an overview of the process:

1. Decide on your *similarity* or *distance* metric.
2. Split the original labeled dataset into training and test data.
3. Pick an evaluation metric. (Misclassification rate is a good one. We'll explain this more in a bit.)
4. Run k-NN a few times, changing *k* and checking the evaluation measure.
5. Optimize *k* by picking the one with the best evaluation measure.
6. Once you've chosen *k*, use the same training set and now create a new test set with the people's ages and incomes that you have *no*

*labels* for, and want to predict. In this case, your new test set only has one lonely row, for the 57-year-old.

## Similarity or distance metrics

Definitions of "closeness" and similarity vary depending on the context: closeness in social networks could be defined as the number of overlapping friends, for example.

For the sake of our problem of what a neighbor is, we can use Euclidean distance on the plane if the variables are on the same scale. And that can sometimes be a big IF.

---

# Caution: Modeling Danger Ahead!

The scalings question is a really big deal, and if you do it wrong, your model could just suck.

Let's consider an example: Say you measure age in years, income in dollars, and credit rating as credit scores normally are given—something like SAT scores. Then two people would be represented by triplets such as $(25, 54000, 700)$ and $(35, 76000, 730)$. In particular, their "distance" would be completely dominated by the difference in their salaries.

On the other hand, if you instead measured salary in *thousands of dollars*, they'd be represented by the triplets $(25, 54, 700)$ and $(35, 76, 730)$, which would give all three variables similar kinds of influence.

Ultimately the way you scale your variables, or equivalently in this situation the way you define your concept of distance, has a potentially enormous effect on the output. In statistics it is called your "prior."

---

Euclidean distance is a good go-to distance metric for attributes that are real-valued and can be plotted on a plane or in multidimensional space. Some others are:

*Cosine Similarity*

Also can be used between two real-valued vectors, $\vec{x}$ and $\vec{y}$, and will yield a value between −1 (exact opposite) and 1 (exactly the same) with 0 in between meaning independent. Recall the definition $\cos\left(\vec{x},\vec{y}\right)=\frac{\vec{x}\cdot\vec{y}}{\|\vec{x}\|\,\|\vec{y}\|}$.

*Jaccard Distance or Similarity*

This gives the distance between a set of objects—for example, a list of Cathy's friends $A=\{Kahn, Mark, Laura, \ldots\}$ and a list of Rachel's friends $B=\{Mladen, Kahn, Mark, \ldots\}$—and says how similar those two sets are: $J(A,B)=\frac{|A\cap B|}{|A\cup B|}$.

*Mahalanobis Distance*

Also can be used between two real-valued vectors and has the advantage over Euclidean distance that it takes into account correlation and is scale-invariant. $d\left(\vec{x},\vec{y}\right)=\sqrt{\left(\vec{x}-\vec{y}\right)^T S^{-1}\left(\vec{x}-\vec{y}\right)}$, where $S$ is the covariance matrix.

*Hamming Distance*

Can be used to find the distance between two strings or pairs of words or DNA sequences of the same length. The distance between olive and ocean is 4 because aside from the "o" the other 4 letters are different. The distance between shoe and hose is 3 because aside from the "e" the other 3 letters are different. You just go through each position and check whether the letters the same in that position, and if not, increment your count by 1.

*Manhattan*

This is also a distance between two real-valued k-dimensional vectors. The image to have in mind is that of a taxi having to travel the city streets of Manhattan, which is laid out in a grid-like fashion (you can't cut diagonally across buildings). The distance is therefore defined as $d\left(\vec{x},\vec{y}\right)=\Sigma_i^k |x_i-y_i|$, where $i$ is the $i$th element of each of the vectors.

There are many more distance metrics available to you depending on your type of data. We start with a Google search when we're not sure where to start.

What if your attributes are a mixture of kinds of data? This happens in the case of the movie ratings example: some were numerical attributes, such as budget and number of actors, and one was categorical, genre. But you can always define your own custom distance metric.

For example, you can say if movies are the same genre, that will contribute "0" to their distance. But if they're of a different genre, that will contribute "10," where you picked the value 10 based on the fact that this was on the same scale as budget (millions of dollars), which is in the range of 0 and 100. You could do the same with number of actors. You could play around with the 10; maybe 50 is better.

You'll want to justify why you're making these choices. The justification could be that you tried different values and when you tested the algorithm, this gave the best evaluation metric. Essentially this 10 is either a second tuning parameter that you've introduced into the algorithm on top of the $k$, or a prior you've put on the model, depending on your point of view and how it's used.

### Training and test sets

For any machine learning algorithm, the general approach is to have a training phase, during which you create a model and "train it"; and then you have a testing phase, where you use new data to test how good the model is.

For k-NN, the training phase is straightforward: it's just reading in your data with the "high" or "low" credit data points marked. In testing, you pretend you don't know the true label and see how good you are at guessing using the k-NN algorithm.

To do this, you'll need to save some clean data from the overall data for the testing phase. Usually you want to save randomly selected data, let's say 20%.

Your R console might look like this:

```
> head(data)
  age income credit
1  69      3    low
2  66     57    low
3  49     79    low
4  49     17    low
5  58     26   high
6  44     71   high

n.points <- 1000 # number of rows in the dataset
sampling.rate <- 0.8
```

```
# we need the number of points in the test set to calculate
# the misclassification rate
num.test.set.labels <- n.points * (1 - sampling.rate)

# randomly sample which rows will go in the training set
training <- sample(1:n.points, sampling.rate * n.points,
                   replace=FALSE)
# define the training set to be those rows
train <- subset(data[training, ], select = c(Age, Income))

# the other rows are going into the test set
testing <- setdiff(1:n.points, training)
# define the test set to be the other rows
test <- subset(data[testing, ], select = c(Age, Income))

# this is the subset of labels for the training set
cl <- data$Credit[training]
# subset of labels for the test set, we're withholding these
true.labels <- data$Credit[testing]
```

## Pick an evaluation metric

How do you evaluate whether your model did a good job?

This isn't easy or universal—you may decide you want to penalize certain kinds of misclassification more than others. False negatives may be way worse than false positives. Coming up with the evaluation metric could be something you work on with a domain expert.

For example, if you were using a classification algorithm to predict whether someone had cancer or not, you would want to minimize false negatives (misdiagnosing someone as not having cancer when they actually do), so you could work with a doctor to tune your evaluation metric.

Note you want to be careful because if you really wanted to have *no* false negatives, you could just tell *everyone* they have cancer. So it's a trade-off between *sensitivity* and *specificity*, where sensitivity is here defined as the probability of correctly diagnosing an ill patient as ill; specificity is here defined as the probability of correctly diagnosing a well patient as well.

**Other Terms for Sensitivity and Specificity**

Sensitivity is also called the *true positive rate* or *recall* and varies based on what academic field you come from, but they all mean the same thing. And *specificity* is also called the *true negative rate*. There is also the *false positive rate* and the *false negative rate*, and these don't get other special names.

Another evaluation metric you could use is *precision*, defined in Chapter 5. The fact that some of the same formulas have different names is due to the fact that different academic disciplines have developed these ideas separately. So *precision* and *recall* are the quantities used in the field of information retrieval. Note that *precision* is not the same thing as *specificity*.

Finally, we have *accuracy*, which is the ratio of the number of correct labels to the total number of labels, and the misclassification rate, which is just 1–*accuracy*. Minimizing the *misclassification rate* then just amounts to maximizing *accuracy*.

## Putting it all together

Now that you have a distance measure and an evaluation metric, you're ready to roll.

For each person in your test set, you'll pretend you don't know his label. Look at the labels of his three nearest neighbors, say, and use the label of the majority vote to label him. You'll label all the members of the test set and then use the misclassification rate to see how well you did. All this is done automatically in R, with just this single line of R code:

```
knn (train, test, cl, k=3)
```

## Choosing k

How do you choose $k$? This is a parameter you have control over. You might need to understand your data pretty well to get a good guess, and then you can try a few different $k$'s and see how your evaluation changes. So you'll run k-nn a few times, changing $k$, and checking the evaluation metric each time.

**Binary Classes**

When you have binary classes like "high credit" or "low credit," picking *k* to be an odd number can be a good idea because there will always be a majority vote, no ties. If there is a tie, the algorithm just randomly picks.

```
# we'll loop through and see what the misclassification rate
# is for different values of k
for (k in 1:20) {
    print(k)
    predicted.labels <- knn(train, test, cl, k)
    # We're using the R function knn()
    num.incorrect.labels <- sum(predicted.labels != true.labels)
    misclassification.rate <- num.incorrect.labels /
                                      num.test.set.labels
    print(misclassification.rate)
}
```

Here's the output in the form (k, misclassification rate):

```
k  misclassification.rate
1, 0.28
2, 0.315
3, 0.26
4, 0.255
5, 0.23
6, 0.26
7, 0.25
8, 0.25
9, 0.235
10, 0.24
```

So let's go with *k* = 5 because it has the lowest misclassification rate, and now you can apply it to your guy who is 57 with a $37,000 salary. In the R console, it looks like:

```
> test <- c(57,37)
> knn(train,test,cl, k = 5)
[1] low
```

The output by majority vote is a low credit score when k = 5.

**Test Set in k-NN**

Notice we used the function knn() twice and used it in different ways. In the first way, the test set was some data we were using to evaluate how good the model was. In the second way, the "test" set was actually a new data point that we wanted a prediction for. We could also have given it many rows of people who we wanted predictions for. But notice that R doesn't know the difference whether what you're putting in for the test set is truly a "test" set where you know the real labels, or a test set where you don't know and want predictions.

## What are the modeling assumptions?

In the previous chapter we discussed modeling and modeling assumptions. So what were the modeling assumptions here?

The k-NN algorithm is an example of a nonparametric approach. You had no modeling assumptions about the underlying data-generating distributions, and you weren't attempting to estimate any parameters. But you still made *some* assumptions, which were:

- Data is in some feature space where a notion of "distance" makes sense.

- Training data has been labeled or classified into two or more classes.

- You pick the number of neighbors to use, $k$.

- You're assuming that the *observed features* and the *labels* are somehow associated. They may not be, but ultimately your evaluation metric will help you determine how good the algorithm is at labeling. You might want to add more features and check how that alters the evaluation metric. You'd then be tuning both *which* features you were using and $k$. But as always, you're in danger here of overfitting.

Both linear regression and k-NN are examples of "supervised learning," where you've observed both $x$ and $y$, and you want to know the function that brings $x$ to $y$. Next up, we'll look at an algorithm you can use when you don't know what the right answer is.

# k-means

So far we've only seen supervised learning, where we know beforehand what label (aka the "right answer") is and we're trying to get our model to be as accurate as possible, defined by our chosen evaluation metric.

k-means is the first *unsupervised* learning technique we'll look into, where the goal of the algorithm is to determine the definition of the right answer by finding clusters of data for you.

Let's say you have some kind of data at the user level, e.g., Google+ data, survey data, medical data, or SAT scores.

Start by adding structure to your data. Namely, assume each row of your dataset corresponds to a user as follows:

```
age gender income state household size
```

Your goal is to *segment* the users. This process is known by various names: besides being called segmenting, you could say that you're going to *stratify*, *group*, or *cluster* the data. They all mean finding similar types of users and bunching them together.

Why would you want to do this? Here are a few examples:

- You might want to give different users different experiences. Marketing often does this; for example, to offer toner to people who are known to own printers.

- You might have a model that works better for specific groups. Or you might have different models for different groups.

- Hierarchical modeling in statistics does something like this; for example, to separately model geographical effects from household effects in survey results.

To see why an algorithm like this might be useful, let's first try to construct something by hand. That might mean you'd bucket users using handmade thresholds.

So for an attribute like age, you'd create bins: 20–24, 25–30, etc. The same technique could be used for other attributes like income. States or cities are in some sense their own buckets, but you might want fewer buckets, depending on your model and the number of data points. In that case, you could bucket the buckets and think of "East Coast" and "Midwest" or something like that.

Say you've done that for each attribute. You may have 10 age buckets, 2 gender buckets, and so on, which would result in $10 \times 2 \times 50 \times 10 \times 3 = 30{,}000$ possible bins, which is big.

Imagine this data existing in a five-dimensional space where each axis corresponds to one attribute. So there's a gender axis, an income axis, and so on. You can also label the various possible buckets along the corresponding axes, and if you did so, the resulting grid would consist of every possible bin—a bin for each possible combination of attributes.

Each user would then live in one of those 30,000 five-dimensional cells. But wait, it's highly unlikely you'd want to build a different marketing campaign for each bin. So you'd have to bin the bins...

Now you likely see the utility of having an algorithm to do this for you, especially if you could choose beforehand how many bins you want. That's exactly what k-means is: a *clustering* algorithm where *k* is the number of bins.

### 2D version

Let's back up to a simpler example than the five-dimensional one we just discussed. Let's say you have users where you know how many ads have been shown to each user (the number of impressions) and how many times each has clicked on an ad (number of clicks).

Figure 3-9 shows a simplistic picture that illustrates what this might look like.

Visually you can see in the top-left that the data naturally falls into clusters. This may be easy for you to do with your eyes when it's only in two dimensions and there aren't that many points, but when you get to higher dimensions and more data, you need an algorithm to help with this pattern-finding process. k-means algorithm looks for clusters in *d* dimensions, where *d* is the number of features for each data point.

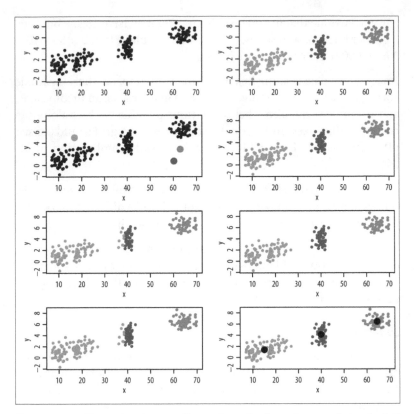

*Figure 3-9. Clustering in two dimensions; look at the panels in the left column from top to bottom, and then the right column from top to bottom*

Here's how the algorithm illustrated in Figure 3-9 works:

1. Initially, you randomly pick $k$ centroids (or points that will be the center of your clusters) in $d$-space. Try to make them near the data but different from one another.

2. Then assign each data point to the closest centroid.

3. Move the centroids to the average location of the data points (which correspond to users in this example) assigned to it.

4. Repeat the preceding two steps until the assignments don't change, or change very little.

It's up to you to interpret if there's a natural way to describe these groups once the algorithm's done. Sometimes you'll need to jiggle around $k$ a few times before you get natural groupings.

This is an example of *unsupervised* learning because the labels are not known and are instead discovered by the algorithm.

k-means has some known issues:

- Choosing $k$ is more an art than a science, although there are bounds: $1 \leq k \leq n$, where $n$ is number of data points.

- There are convergence issues—the solution can fail to exist, if the algorithm falls into a loop, for example, and keeps going back and forth between two possible solutions, or in other words, there isn't a single unique solution.

- Interpretability can be a problem—sometimes the answer isn't at all useful. Indeed that's often the biggest problem.

In spite of these issues, it's pretty fast (compared to other clustering algorithms), and there are broad applications in marketing, computer vision (partitioning an image), or as a starting point for other models.

In practice, this is just one line of code in R:

```
kmeans(x, centers, iter.max = 10, nstart = 1,
        algorithm = c("Hartigan-Wong", "Lloyd", "Forgy",
                    "MacQueen"))
```

Your dataset needs to be a matrix, x, each column of which is one of your features. You specify $k$ by selecting centers. It defaults to a certain number of iterations, which is an argument you can change. You can also select the specific algorithm it uses to discover the clusters.

## Historical Perspective: k-means

Wait, didn't we just describe the algorithm? It turns out there's more than one way to go after k-means clustering.

The standard k-means algorithm is attributed to separate work by Hugo Steinhaus and Stuart Lloyd in 1957, but it wasn't called "k-means" then. The first person to use that term was James MacQueen in 1967. It wasn't published outside Bell Labs until 1982.

Newer versions of the algorithm are Hartigan-Wong and Lloyd and Forgy, named for their inventors and developed throughout the '60s and '70s. The algorithm we described is the default, Hartigan-Wong. It's fine to use the default.

As history keeps marching on, it's worth checking out the more recent k-means++ developed in 2007 by David Arthur and Sergei Vassilvitskii (now at Google), which helps avoid convergence issues with k-means by optimizing the initial seeds.

# Exercise: Basic Machine Learning Algorithms

Continue with the NYC (Manhattan) Housing dataset you worked with in the preceding chapter: *http://abt.cm/1g3A12P*.

- Analyze sales using regression with any predictors you feel are relevant. Justify why regression was appropriate to use.
- Visualize the coefficients and fitted model.
- Predict the neighborhood using a k-NN classifier. Be sure to withhold a subset of the data for testing. Find the variables and the *k* that give you the lowest prediction error.
- Report and visualize your findings.
- Describe any decisions that could be made or actions that could be taken from this analysis.

## Solutions

In the preceding chapter, we showed how to explore and clean this dataset, so you'll want to do that first before you build your regression model. Following are two pieces of R code. The first shows how you might go about building your regression models, and the second shows how you might clean and prepare your data and then build a k-NN classifier.

### Sample R code: Linear regression on the housing dataset

```
Author: Ben Reddy

model1 <- lm(log(sale.price.n) ~ log(gross.sqft),data=bk.homes)
## what's going on here?

bk.homes[which(bk.homes$gross.sqft==0),]

bk.homes <- bk.homes[which(bk.homes$gross.sqft>0 &
```

```
            bk.homes$land.sqft>0),]
model1 <- lm(log(sale.price.n) ~ log(gross.sqft),data=bk.homes)
summary(model1)

plot(log(bk.homes$gross.sqft),log(bk.homes$sale.price.n))
abline(model1,col="red",lwd=2)
plot(resid(model1))

model2 <- lm(log(sale.price.n) ~ log(gross.sqft) +
  log(land.sqft) + factor(neighborhood),data=bk.homes)
summary(model2)
plot(resid(model2))

## leave out intercept for ease of interpretability
model2a <- lm(log(sale.price.n) ~ 0 + log(gross.sqft) +
  log(land.sqft) + factor(neighborhood),data=bk.homes)
summary(model2a)
plot(resid(model2a))

## add building type
model3 <- lm(log(sale.price.n) ~ log(gross.sqft) +
  log(land.sqft) + factor(neighborhood) +
  factor(building.class.category),data=bk.homes)
summary(model3)
plot(resid(model3))

## interact neighborhood and building type
model4 <- lm(log(sale.price.n) ~ log(gross.sqft) +
  log(land.sqft) +  factor(neighborhood)*
  factor(building.class.category),data=bk.homes)
summary(model4)
plot(resid(model4))
```

## Sample R code: K-NN on the housing dataset

```
Author: Ben Reddy
require(gdata)
require(geoPlot)
require(class)

mt <- read.xls("rollingsales_manhattan.xls",
  pattern="BOROUGH",stringsAsFactors=FALSE)
head(mt)
summary(mt)

names(mt) <- tolower(names(mt))

mt$sale.price.n <- as.numeric(gsub("[^[:digit:]]","",
                                mt$sale.price))
sum(is.na(mt$sale.price.n))
sum(mt$sale.price.n==0)
```

```
names(mt) <- tolower(names(mt))

## clean/format the data with regular expressions
mt$gross.sqft <- as.numeric(gsub("[^[:digit:]]","",
                            mt$gross.square.feet))
mt$land.sqft <- as.numeric(gsub("[^[:digit:]]","",
                            mt$land.square.feet))

mt$sale.date <- as.Date(mt$sale.date)
mt$year.built <- as.numeric(as.character(mt$year.built))
mt$zip.code <- as.character(mt$zip.code)

## - standardize data (set year built start to 0; land and
gross sq ft; sale price (exclude $0 and possibly others); possi
bly tax block; outside dataset for coords of tax block/lot?)
min_price <- 10000
mt <- mt[which(mt$sale.price.n>=min_price),]

n_obs <- dim(mt)[1]

mt$address.noapt <- gsub("[,][[:print:]]*","",
                        gsub("[ ]+"," ",trim(mt$address)))

mt_add <- unique(data.frame(mt$address.noapt,mt$zip.code,
                    stringsAsFactors=FALSE))
names(mt_add) <- c("address.noapt","zip.code")
mt_add <- mt_add[order(mt_add$address.noapt),]

#find duplicate addresses with different zip codes
dup <- duplicated(mt_add$address.noapt)
# remove them
dup_add <- mt_add[dup,1]
mt_add <- mt_add[(mt_add$address.noapt != dup_add[1] &
        mt_add$address.noapt != dup_add[2]),]

n_add <- dim(mt_add)[1]

# sample 500 addresses so we don't run over our Google Maps
API daily limit (and so we're not waiting forever)
n_sample <- 500
add_sample <- mt_add[sample.int(n_add,size=n_sample),]

# first, try a query with the addresses we have
query_list <- addrListLookup(data.frame(1:n_sample,
  add_sample$address.noapt,rep("NEW YORK",times=n_sample),
  rep("NY",times=n_sample),add_sample$zip.code,
  rep("US",times=n_sample)))[,1:4]

query_list$matched <- (query_list$latitude != 0)
```

```
unmatched_inds <- which(!query_list$matched)
unmatched <- length(unmatched_inds)

# try changing EAST/WEST to E/W
query_list[unmatched_inds,1:4] <- addrListLookup
  (data.frame(1:unmatched,gsub(" WEST "," W ",
  gsub(" EAST "," E ",add_sample[unmatched_inds,1])),
  rep("NEW YORK",times=unmatched), rep("NY",times=unmatched),
      add_sample[unmatched_inds,2],rep("US",times=unmatched)))[,
1:4]

query_list$matched <- (query_list$latitude != 0)
unmatched_inds <- which(!query_list$matched)
unmatched <- length(unmatched_inds)

# try changing STREET/AVENUE to ST/AVE
query_list[unmatched_inds,1:4] <- addrListLookup
  (data.frame(1:unmatched,gsub(" WEST "," W ",
  gsub(" EAST "," E ",gsub(" STREET"," ST",
  gsub(" AVENUE"," AVE",add_sample[unmatched_inds,1])))),
  rep("NEW YORK",times=unmatched), rep("NY",times=unmatched),
      add_sample[unmatched_inds,2],rep("US",times=unmatched)))[,
1:4]

query_list$matched <- (query_list$latitude != 0)
unmatched_inds <- which(!query_list$matched)
unmatched <- length(unmatched_inds)

## have to be satisfied for now
add_sample <- cbind(add_sample,query_list$latitude,
  query_list$longitude)
names(add_sample)[3:4] <- c("latitude","longitude")

add_sample <- add_sample[add_sample$latitude!=0,]

add_use <- merge(mt,add_sample)
add_use <- add_use[!is.na(add_use$latitude),]

# map coordinates
map_coords <- add_use[,c(2,4,26,27)]
table(map_coords$neighborhood)
map_coords$neighborhood <- as.factor(map_coords$neighborhood)

geoPlot(map_coords,zoom=12,color=map_coords$neighborhood)


## - knn function
## - there are more efficient ways of doing this,
## but oh well...

map_coords$class <- as.numeric(map_coords$neighborhood)
```

```
n_cases <- dim(map_coords)[1]
split <- 0.8

train_inds <- sample.int(n_cases,floor(split*n_cases))
test_inds <- (1:n_cases)[-train_inds]

k_max <- 10
knn_pred <- matrix(NA,ncol=k_max,nrow=length(test_inds))
knn_test_error <- rep(NA,times=k_max)

for (i in 1:k_max) {
    knn_pred[,i] <- knn(map_coords[train_inds,3:4],
  map_coords[test_inds,3:4],cl=map_coords[train_inds,5],k=i)
    knn_test_error[i] <- sum(knn_pred[,i]!=
      map_coords[test_inds,5])/length(test_inds)
}

plot(1:k_max,knn_test_error)
```

## Modeling and Algorithms at Scale

The data you've been dealing with so far in this chapter has been pretty small on the Big Data spectrum. What happens to these models and algorithms when you have to scale up to massive datasets?

In some cases, it's entirely appropriate to sample and work with a smaller dataset, or to run the same model across multiple *sharded* datasets. (Sharding is where the data is broken up into pieces and divided among diffrent machines, and then you look at the empirical distribution of the estimators across models.) In other words, there are statistical solutions to these engineering challenges.

However, in some cases we want to fit these models at scale, and the challenge of scaling up models generally translates to the challenge of creating parallelized versions or approximations of the optimization methods. Linear regression at scale, for example, relies on matrix inversions or approximations of matrix inversions.

Optimization with Big Data calls for new approaches and theory—this is the frontier! From a 2013 talk by Peter Richtarik from the University of Edinburgh: "In the Big Data domain classical approaches that rely on optimization methods with multiple iterations are not applicable as the computational cost of even a single iteration is often too excessive; these methods were developed in the past when problems of huge sizes were rare to find. We thus need new methods which would be simple, gentle with data handling and memory requirements, and scalable. Our ability to solve truly huge scale problems

goes hand in hand with our ability to utilize modern parallel computing architectures such as multicore processors, graphical processing units, and computer clusters."

Much of this is outside the scope of the book, but a data scientist needs
to be aware of these issues, and some of this is discussed in Chapter 14.

# Summing It All Up

We've now introduced you to three algorithms that are the basis for
the solutions to many real-world problems. If you understand these
three, you're already in good shape. If you don't, don't worry, it takes
a while to sink in.

Regression is the basis of many forecasting and classification or prediction models in a variety of contexts. We showed you how you can
predict a continuous outcome variable with one or more predictors.
We'll revisit it again in Chapter 5, where we'll learn *logistic* regression,
which can be used for classification of binary outcomes; and in Chapter 6, where we see it in the context of time series modeling. We'll also
build up your feature selection skills in Chapter 7.

k-NN and k-means are two examples of clustering algorithms, where
we want to group together similar objects. Here the notions of *distance*
and *evaluation measures* became important, and we saw there is some
subjectivity involved in picking these. We'll explore clustering algorithms including Naive Bayes in the next chapter, and in the context
of social networks (Chapter 10). As we'll see, *graph clustering* is an
interesting area of research. Other examples of clustering algorithms
not explored in this book are *hierarchical clustering* and *model-based
clustering*.

For further reading and a more advanced treatment of this material,
we recommend the standard classic Hastie and Tibshirani book, *Elements of Statistical Learning (http://stanford.io/16hcTKn)* (Springer).
For an in-depth exploration of building regression models in a Bayesian context, we highly recommend Andrew Gelman and Jennifer
Hill's *Data Analysis using Regression and Multilevel/Hierarchical
Models*.

# Thought Experiment: Automated Statistician

Rachel attended a workshop in Big Data Mining at Imperial College London in May 2013. One of the speakers, Professor Zoubin Ghahramani from Cambridge University, said that one of his long-term research projects was to build an "automated statistician." What do you think that means? What do you think would go into building one?

Does the idea scare you? Should it?

CHAPTER 4

# Spam Filters, Naive Bayes, and Wrangling

The contributor for this chapter is Jake Hofman. Jake is at Microsoft Research after recently leaving Yahoo! Research. He got a PhD in physics at Columbia and regularly teaches a fantastic course on data-driven modeling at Columbia, as well as a newer course in computational social science.

As with our other presenters, we first took a look at Jake's data science profile. It turns out he is an expert on a category that he added to the data science profile called "data wrangling." He confessed that he doesn't know if he spends so much time on it because he's good at it or because he's bad at it. (He's good at it.)

## Thought Experiment: Learning by Example

Let's start by looking at a bunch of text shown in Figure 4-1, whose rows seem to contain the subject and first line of an email in an inbox.

You may notice that several of the rows of text look like spam.

How did you figure this out? Can you write code to automate the spam filter that your brain represents?

| | | |
|---|---|---|
| Pure Saffron Extract | Melt Fat Away - Drop 11-lbs in 7 Days! - Melt Fat Away - Drop 11-lbs in 7 Days! Melt Fat Away - Drop 11-lbs i |
| Blue Sky Auto | Car Loans Available - Bad Credit Accepted |
| Watch The Video | Shocking Discovery Gets You Laid - Scientists at Harvard University have discovered a strange secret that allo |
| Casino | Casino Promotions - With the Slots of Vegas Instant-Win Scratch Ticket Game you can get $100 on the hous |
| Designer Watch Replica | Replica Watches On Sale - Replica Watches: Swiss Luxury Watch Replicas, Rolex, Omega, Breitling Check |
| A.C., me (10) | I'm late to this party - I'm free and interested. Tell me more! I'd have to think about the students, but I know so |
| Rachel .. Christoforos (18) | Fwd: Invitation to speak at upcoming Big Data Workshop, hosted by Imperial College London - Dear Rachel, I! |
| Fat Burning Hormone | 17 Foods that GET RID of stomach fat |
| Kaplan University | Kaplan University online and campus degree programs |
| Dinn Trophy | Sport Plaques - As Low As $4.29 - View this message in a browser. Shop Sport Plaques Shop Now> Change |
| me, Philipp (2) | checking in - Hi Rachel, I know! I had started writing a few emails to you, but then I (obviously) didn't sent |
| me, Matthew (3) | doing data science - Hi Matt, Not a duplicate (just FYI if that helps debug) Well, so the status is that we're in t |
| Luxury Replicas | Rolex, Breitling, Chanel, Omega, LV, and muchMore! - Super Replicas - Luxury Watches, Bags, Jewelry, Pho |
| Watch this video and wom. | Watch this video and women will adore you - Can you get laid using just the words in this video? Click Here To |
| Adriana | I ADDED YOU to my Private Wish List - Sorry, I've been out of town but I am back and I'm looking for a good ti |

*Figure 4-1. Suspiciously spammy*

Rachel's class had a few ideas about what things might be clear signs of spam:

- Any email is spam if it contains Viagra references. That's a good rule to start with, but as you've likely seen in your own email, people figured out this spam filter rule and got around it by modifying the spelling. (It's sad that spammers are so smart and aren't working on more important projects than selling lots of Viagra…)

- Maybe something about the length of the subject gives it away as spam, or perhaps excessive use of exclamation points or other punctuation. But some words like "Yahoo!" are authentic, so you don't want to make your rule too simplistic.

And here are a few suggestions regarding code you could write to identify spam:

- Try a probabilistic model. In other words, should you not have simple rules, but have many rules of thumb that aggregate together to provide the probability of a given email being spam? This is a great idea.

- What about k-nearest neighbors or linear regression? You learned about these techniques in the previous chapter, but do they apply to this kind of problem? (Hint: the answer is "No.")

In this chapter, we'll use Naive Bayes to solve this problem, which is in some sense in between the two. But first…

# Why Won't Linear Regression Work for Filtering Spam?

Because we're already familiar with linear regression and that's a tool in our toolbelt, let's start by talking through what we'd need to do in order to try to use linear regression. We already know that's *not* what we're going to use, but let's talk it through to get to why. Imagine a dataset or matrix where each row represents a different email message (it could be keyed by email_id). Now let's make each word in the email a *feature*—this means that we create a column called "Viagra," for example, and then for any message that has the word Viagra in it at least once, we put a 1 in; otherwise we assign a 0. Alternatively, we could put the number of times the word appears. Then each column represents the appearance of a different word.

Thinking back to the previous chapter, in order to use linear regression, we need to have a training set of emails where the messages have *already* been labeled with some outcome variable. In this case, the outcomes are either spam or not. We could do this by having human evaluators label messages "spam," which is a reasonable, albeit time-intensive, solution. Another way to do it would be to take an existing spam filter such as Gmail's spam filter and use those labels. (Now of course if you already had a Gmail spam filter, it's hard to understand why you might also want to build another spam filter in the first place, but let's just say you do.) Once you build a model, email messages would come in without a label, and you'd use your model to predict the labels.

The first thing to consider is that your target is binary (0 if not spam, 1 if spam)—you wouldn't get a 0 or a 1 using linear regression; you'd get a number. Strictly speaking, this option really isn't ideal; linear regression is aimed at modeling a continuous output and this is binary.

This issue is basically a nonstarter. We should use a model appropriate for the data. But if we wanted to fit it in R, in theory it could still work. R doesn't check for us whether the model is appropriate or not. We could go for it, fit a linear model, and then use that to predict and then choose a critical value so that above that predicted value we call it "1" and below we call it "0."

But if we went ahead and tried, it still wouldn't work because there are too many variables compared to observations! We have on the order of 10,000 emails with on the order of 100,000 words. This won't work. Technically, this corresponds to the fact that the matrix in the equation for linear regression is not invertible—in fact, it's not even close. Moreover, maybe we can't even store it because it's so huge.

Maybe we could limit it to the top 10,000 words? Then we could at least have an invertible matrix. Even so, that's too many variables versus observations to feel good about it. With carefully chosen feature selection and domain expertise, we could limit it to 100 words and that could be enough! But again, we'd still have the issue that linear regression is not the appropriate model for a binary outcome.

**Aside: State of the Art for Spam Filters**
In the last five years, people have started using stochastic gradient methods to avoid the noninvertible (overfitting) matrix problem. Switching to logistic regression with stochastic gradient methods helped a lot, and can account for correlations between words. Even so, Naive Bayes is pretty impressively good considering how simple it is.

## How About k-nearest Neighbors?

We're going to get to Naive Bayes shortly, we promise, but let's take a minute to think about trying to use k-nearest neighbors (k-NN) to create a spam filter. We would still need to choose features, probably corresponding to words, and we'd likely define the value of those features to be 0 or 1, depending on whether the word is present or not. Then, we'd need to define when two emails are "near" each other based on which words they both contain.

Again, with 10,000 emails and 100,000 words, we'll encounter a problem, different from the noninvertible matrix problem. Namely, the space we'd be working in has *too many dimensions*. Yes, computing distances in a 100,000-dimensional space requires lots of computational work. But that's not the real problem.

The real problem is even more basic: even our nearest neighbors are really far away. This is called "the curse of dimensionality," and it makes k-NN a poor algorithm in this case.

# Aside: Digit Recognition

Say you want an algorithm to recognize pictures of hand-written digits as shown in Figure 4-2. In this case, k-NN works well.

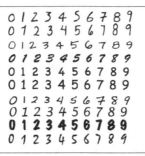

*Figure 4-2. Handwritten digits*

To set it up, you take your underlying representation apart pixel by pixel—say in a 16x16 grid of pixels—and measure how bright each pixel is. Unwrap the 16x16 grid and put it into a 256-dimensional space, which has a natural archimedean metric. That is to say, the distance between two different points on this space is the square root of the sum of the squares of the differences between their entries. In other words, it's the length of the vector going from one point to the other or vice versa. Then you apply the k-NN algorithm.

If you vary the number of neighbors, it changes the shape of the boundary, and you can tune $k$ to prevent overfitting. If you're careful, you can get 97% accuracy with a sufficiently large dataset.

Moreover, the result can be viewed in a "confusion matrix." A *confusion matrix* is used when you are trying to classify objects into $k$ bins, and is a $k \times k$ matrix corresponding to actual label versus predicted label, and the $(i, j)$th element of the matrix is a count of the number of items that were actually labeled $i$ that were predicted to have label $j$. From a confusion matrix, you can get *accuracy*, the proportion of total predictions that were correct. In the previous chapter, we discussed the misclassification rate. Notice that accuracy = 1 - misclassification rate.

# Naive Bayes

So are we at a loss now that two methods we're familiar with, linear regression and k-NN, won't work for the spam filter problem? No! Naive Bayes is another classification method at our disposal that scales well and has nice intuitive appeal.

## Bayes Law

Let's start with an even simpler example than the spam filter to get a feel for how Naive Bayes works. Let's say we're testing for a rare disease, where 1% of the population is infected. We have a highly sensitive and specific test, which is not quite perfect:

- 99% of sick patients test positive.
- 99% of healthy patients test negative.

Given that a patient tests positive, what is the probability that the patient is actually sick?

A naive approach to answering this question is this: Imagine we have $100 \times 100 = 10,000$ perfectly representative people. That would mean that 100 are sick, and 9,900 are healthy. Moreover, after giving all of them the test we'd get 99 sick people testing sick, but 99 healthy people testing sick as well. If you test positive, in other words, you're equally likely to be healthy or sick; the answer is 50%. A tree diagram of this approach is shown in Figure 4-3.

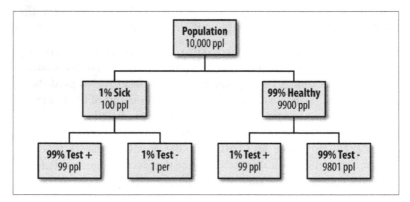

*Figure 4-3. Tree diagram to build intuition*

Let's do it again using fancy notation so we'll feel smart.

Recall from your basic statistics course that, given events $x$ and $y$, there's a relationship between the probabilities of either event (denoted $p(x)$ and $p(y)$), the joint probabilities (both happen, which is denoted $p(x, y)$), and conditional probabilities (event $x$ happens given $y$ happens, denoted $p(x|y)$) as follows:

$$p(y|x)p(x) = p(x, y) = p(x|y)p(y)$$

Using that, we solve for $p(y|x)$ (assuming $p(x) \neq 0$) to get what is called *Bayes' Law*:

$$p(y|x) = \frac{p(x|y)p(y)}{p(x)}$$

The denominator term, $p(x)$, is often implicitly computed and can thus be treated as a "normalization constant." In our current situation, set $y$ to refer to the event "I am sick," or "sick" for shorthand; and set $x$ to refer to the event "the test is positive," or "+" for shorthand. Then we actually know, or at least can compute, every term:

$$p(sick|+) = \frac{p(+|sick)p(sick)}{p(+)} = \frac{0.99 \cdot 0.01}{0.99 \cdot 0.01 + 0.01 \cdot 0.99} = 0.50 = 50\%$$

## A Spam Filter for Individual Words

So how do we use Bayes' Law to create a good spam filter? Think about it this way: if the word "Viagra" appears, this adds to the probability that the email is spam. But it's not conclusive, yet. We need to see what else is in the email.

Let's first focus on just one word at a time, which we generically call "word." Then, applying Bayes' Law, we have:

$$p(spam|word) = \frac{p(word|spam)p(spam)}{p(word)}$$

The righthand side of this equation is computable using enough prelabeled data. If we refer to nonspam as "ham" then we only need compute $p(word|spam)$, $p(word|ham)$, $p(spam)$, and $p(ham) = 1-p(spam)$, because we can work out the denominator using the formula we used earlier in our medical test example, namely:

$$p(word) = p(word|spam)p(spam) + p(word|ham)p(ham)$$

In other words, we've boiled it down to a counting exercise: $p(spam)$ counts spam emails versus all emails, $p(word|spam)$ counts the prevalence of those spam emails that contain "word," and $p(word|ham)$ counts the prevalence of the ham emails that contain "word."

To do this yourself, go online and download Enron emails (*https://www.cs.cmu.edu/~enron/*). Let's build a spam filter on that dataset. This really this means we're building a new spam filter on top of the spam filter that existed for the employees of Enron. We'll use their definition of spam to train our spam filter. (This does mean that if the spammers have learned anything since 2001, we're out of luck.)

We could write a quick-and-dirty shell script in bash that runs this, which Jake did. It downloads and unzips the file and creates a folder; each text file is an email; spam and ham go in separate folders.

Let's look at some basic statistics on a random Enron employee's email. We can count 1,500 spam versus 3,672 ham, so we already know $p(spam)$ and $p(ham)$. Using command-line tools, we can also count the number of instances of the word "meeting" in the spam folder:

```
grep -il meeting enron1/spam/*.txt | wc -l
```

This gives 16. Do the same for his ham folder, and we get 153. We can now compute the chance that an email is spam only knowing it contains the word "meeting":

$$\hat{p}(spam) = 1500 / (1500 + 3672) = .29$$

$$\hat{p}(ham) = .71$$

$$\hat{p}(meeting|spam) = 16 / 1500 = .0106$$

$$\hat{p}(meeting|ham) = 153 / 3672 = .0416$$

$$\hat{p}(spam|meeting) = \hat{p}(meeting|spam) * \hat{p}(spam) / \hat{p}(meeting) =$$
$$(.0106 * .29) / (.0106 * .29 + .0416 * .71) = 0.09 = 9\%$$

Take note that we didn't need a fancy programming environment to get this done.

---

Next, we can try:

- "money": 80% chance of being spam
- "viagra": 100% chance
- "enron": 0% chance

This illustrates that the model, as it stands, is overfitting; we are getting overconfident because of biased data. Is it really a slam-dunk that any email containing the word "Viagra" is spam? It's of course possible to write a nonspam email with the word "Viagra," as well as a spam email with the word "Enron."

## A Spam Filter That Combines Words: Naive Bayes

Next, let's do it for all the words. Each email can be represented by a binary vector, whose $j$th entry is 1 or 0 depending on whether the $j$th word appears. Note this is a huge-ass vector, considering how many words we have, and we'd probably want to represent it with the indices of the words that actually show up.

The model's output is the probability that we'd see a given word vector given that we know it's spam (or that it's ham). Denote the email vector to be $x$ and the various entries $x_j$, where the $j$ indexes the words. For now we can denote "is spam" by $c$, and we have the following model for $p(x|c)$, i.e., the probability that the email's vector looks like this considering it's spam:

$$p(x|c) = \prod_j \theta_{jc}{}^{x_j} (1 - \theta_{jc})^{(1-x_j)}$$

The $\theta$ here is the probability that an individual word is present in a spam email. We saw how to compute that in the previous section via counting, and so we can assume we've separately and parallelly computed that for every word.

We are modeling the words *independently* (also known as "independent trials"), which is why we take the product on the righthand side of the preceding formula and don't count how many times they are present. That's why this is called "naive," because we know that there are actually certain words that tend to appear together, and we're ignoring this.

So back to the equation, it's a standard trick when we're dealing with a product of probabilities to take the log of both sides to get a summation instead:

$$log(p(x|c)) = \sum_j x_j log(\theta_j / (1 - \theta_j)) + \sum_j log(1 - \theta_j)$$

 It's helpful to take the log because multiplying together tiny numbers can give us numerical problems.

The term $log(\theta_j / (1 - \theta_j))$ doesn't depend on a given email, just the word, so let's rename it $w_j$ and assume we've computed it once and stored it. Same with the quantity $\sum_j log(1 - \theta_j) = w_0$. Now we have:

$$log(p(x|c)) = \sum_j x_j w_j + w_0$$

The weights that vary by email are the $x_j$s. We need to compute them separately for each email, but that shouldn't be too hard.

We can put together what we know to compute $p(x|c)$, and then use Bayes' Law to get an estimate of $p(c|x)$, which is what we actually want —the other terms in Bayes' Law are easier than this and don't require separate calculations per email. We can also get away with not computing all the terms if we only care whether it's more likely to be spam or to be ham. Then only the varying term needs to be computed.

You may notice that this ends up looking like a linear regression, but instead of computing the coefficients $w_j$ by inverting a huge matrix, the weights come from the Naive Bayes' algorithm.

This algorithm works pretty well, and it's "cheap" to train if we have a prelabeled dataset to train on. Given a ton of emails, we're more or less just counting the words in spam and nonspam emails. If we get more training data, we can easily increment our counts to improve our filter. In practice, there's a global model, which we personalize to individuals. Moreover, there are lots of hardcoded, cheap rules before an email gets put into a fancy and slow model.

Here are some references for more about Bayes' Law:

- "Idiot's Bayes - not so stupid after all?" (*http://goo.gl/sD86Oj*) (The whole paper is about why it doesn't suck, which is related to redundancies in language.)
- "Naive Bayes at Forty: The Independence Assumption in Information" (*http://goo.gl/JqjRg3*)
- "Spam Filtering with Naive Bayes - Which Naive Bayes?" (*http://goo.gl/74CxlZ*)

# Fancy It Up: Laplace Smoothing

Remember the $\theta_j$ from the previous section? That referred to the probability of seeing a given word (indexed by $j$) in a spam email. If you think about it, this is just a ratio of counts: $\theta_j = n_{js} / n_{jc}$, where $n_{js}$ denotes the number of times that word appears in a spam email and $n_{jc}$ denotes the number of times that word appears in any email.

*Laplace Smoothing* refers to the idea of replacing our straight-up estimate of $\theta_j$ with something a bit fancier:

$$\theta_{jc} = (n_{js} + \alpha) / (n_{jc} + \beta)$$

We might fix $\alpha = 1$ and $\beta = 10$, for example, to prevent the possibility of getting 0 or 1 for a probability, which we saw earlier happening with "viagra." Does this seem totally ad hoc? Well, if we want to get fancy, we can see this as equivalent to having a prior and performing a maximal likelihood estimate. Let's get fancy! If we denote by *ML* the maximal likelihood estimate, and by $D$ the dataset, then we have:

$$\theta_{ML} = argmax_\theta p(D | \theta)$$

In other words, the vector of values $\theta_j$ is the answer to the question: for what value of $\theta$ were the data D most probable? If we assume independent trials again, as we did in our first attempt at Naive Bayes, then we want to choose the $\theta_j$ to separately maximize the following quantity for each $j$:

$$log \left( \theta_j^{n_{js}} (1 - \theta_j)^{n_{jc} - n_{js}} \right)$$

If we take the derivative, and set it to zero, we get:

$$\hat{\theta}_j = n_{js} / n_{jc}$$

In other words, just what we had before. So what we've found is that the maximal likelihood estimate recovers your result, as long as we assume independence.

Now let's add a prior. For this discussion we can suppress the $j$ from the notation for clarity, but keep in mind that we are fixing the $j$th word to work on. Denote by *MAP* the *maximum a posteriori* likelihood:

$$\theta_{MAP} = argmax \ p(\theta|D)$$

This similarly answers the question: given the data I saw, which parameter $\theta$ is the most likely?

Here we will apply the spirit of Bayes's Law to transform $\theta_{MAP}$ to get something that is, up to a constant, equivalent to $p(D|\theta) \cdot p(\theta)$. The term $p(\theta)$ is referred to as the "prior," and we have to make an assumption about its form to make this useful. If we make the assumption that the probability distribution of $\theta$ is of the form $\theta^\alpha (1-\theta)^\beta$, for some $\alpha$ and $\beta$, then we recover the Laplace Smoothed result.

---

## Is That a Reasonable Assumption?

Recall that $\theta$ is the chance that a word is in spam if that word is in some email. On the one hand, as long as both $\alpha > 0$ and $\beta > 0$, this distribution vanishes at both 0 and 1. This is reasonable: you want very few words to be expected to *never* appear in spam or to *always* appear in spam.

On the other hand, when $\alpha$ and $\beta$ are large, the shape of the distribution is bunched in the middle, which reflects the prior that most words are equally likely to appear in spam or outside spam. That doesn't seem true either.

A compromise would have $\alpha$ and $\beta$ be positive but small, like 1/5. That would keep your spam filter from being too overzealous without having the wrong idea. Of course, you could relax this prior as you have more and better data; in general, strong priors are only needed when you don't have sufficient data.

---

# Comparing Naive Bayes to k-NN

Sometimes $\alpha$ and $\beta$ are called "pseudocounts." Another common name is "hyperparameters." They're fancy but also simple. It's up to you, the data scientist, to set the values of these two hyperparameters in the numerator and denominator for smoothing, and it gives you two knobs to tune. By contrast, k-NN has one knob, namely $k$, the number of neighbors. Naive Bayes is a linear classifier, while k-NN is not. The curse of dimensionality and large feature sets are a problem for k-NN, while Naive Bayes performs well. k-NN requires no training (just load in the dataset), whereas Naive Bayes does. Both are examples of supervised learning (the data comes labeled).

# Sample Code in bash

```bash
#!/bin/bash
#
# file: enron_naive_bayes.sh
#
# description: trains a simple one-word naive bayes spam
# filter using enron email data
#
# usage: ./enron_naive_bayes.sh <word>
#
# requirements:
#   wget
#
# author: jake hofman (gmail: jhofman)
#

# how to use the code
if [ $# -eq 1 ]
    then
    word=$1
else
    echo "usage: enron_naive_bayes.sh <word>"
    exit
fi

# if the file doesn't exist, download from the web
if ! [ -e enron1.tar.gz ]
    then
    wget 'http://www.aueb.gr/users/ion/data/
    enron-spam/preprocessed/enron1.tar.gz'
fi

# if the directory doesn't exist, uncompress the .tar.gz
if ! [ -d enron1 ]
```

```
        then
            tar zxvf enron1.tar.gz
    fi

    # change into enron1
    cd enron1

    # get counts of total spam, ham, and overall msgs
    Nspam=`ls -l spam/*.txt | wc -l`
    Nham=`ls -l ham/*.txt | wc -l`
    Ntot=$Nspam+$Nham

    echo $Nspam spam examples
    echo $Nham ham examples

    # get counts containing word in spam and ham classes
    Nword_spam=`grep -il $word spam/*.txt | wc -l`
    Nword_ham=`grep -il $word ham/*.txt | wc -l`

    echo $Nword_spam "spam examples containing $word"
    echo $Nword_ham "ham examples containing $word"

    # calculate probabilities using bash calculator "bc"
    Pspam=`echo "scale=4; $Nspam / ($Nspam+$Nham)" | bc`
    Pham=`echo "scale=4; 1-$Pspam" | bc`
    echo
    echo "estimated P(spam) =" $Pspam
    echo "estimated P(ham) =" $Pham

    Pword_spam=`echo "scale=4; $Nword_spam / $Nspam" | bc`
    Pword_ham=`echo "scale=4; $Nword_ham / $Nham" | bc`
    echo "estimated P($word|spam) =" $Pword_spam
    echo "estimated P($word|ham) =" $Pword_ham

    Pspam_word=`echo "scale=4; $Pword_spam*$Pspam" | bc`
    Pham_word=`echo "scale=4; $Pword_ham*$Pham" | bc`
    Pword=`echo "scale=4; $Pspam_word+$Pham_word" | bc`
    Pspam_word=`echo "scale=4; $Pspam_word / $Pword" | bc`
    echo
    echo "P(spam|$word) =" $Pspam_word

    # return original directory
    cd ..
```

# Scraping the Web: APIs and Other Tools

As a data scientist, you're not always just handed some data and asked to go figure something out based on it. Often, you have to actually figure out how to go get some data you need to ask a question, solve a problem, do some research, etc. One way you can do this is with an API. For the sake of this discussion, an API (application programming interface) is something websites provide to developers so they can download data from the website easily and in standard format. (APIs are used for much more than this, but for your purpose this is how you'd typically interact with one.) Usually the developer has to register and receive a "key," which is something like a password. For example, the *New York Times* has an API here (*http://developer.nytimes.com/docs*).

**Warning about APIs**

Always check the terms and services of a website's API before scraping. Additionally, some websites limit what data you have access to through their APIs or how often you can ask for data without paying for it.

When you go this route, you often get back weird formats, sometimes in JSON, but there's no standardization to this standardization; i.e., different websites give you different "standard" formats.

One way to get beyond this is to use Yahoo's YQL language (*http://developer.yahoo.com/yql/*), which allows you to go to the Yahoo! Developer Network and write SQL-like queries that interact with many of the APIs on common sites like this:

```
select * from flickr.photos.search where text="Cat"
and api_key="lksdjflskjdfsldkfj" limit 10
```

The output is standard, and you only have to parse this in Python once.

But what if you want data when there's no API available?

In this case you might want to use something like the Firebug extension for Firefox. You can "inspect the element" on any web page, and Firebug allows you to grab the field inside the HTML. In fact, it gives you access to the full HTML document so you can interact and edit. In this way you can see the HTML as a map of the page and Firebug as a kind of tour guide.

After locating the stuff you want inside the HTML, you can use curl, wget, grep, awk, perl, etc., to write a quick-and-dirty shell script to grab what you want, especially for a one-off grab. If you want to be more systematic, you can also do this using Python or R.

Other parsing tools you might want to look into include:

*lynx and lynx --dump (http://lynx.browser.org/)*
  Good if you pine for the 1970s. Oh wait, 1992. Whatever.

*Beautiful Soup (http://www.crummy.com/software/BeautifulSoup/)*
  Robust but kind of slow.

*Mechanize (http://mechanize.rubyforge.org/Mechanize.html) (or here (http://pypi.python.org/pypi/mechanize/))*
  Super cool as well, but it doesn't parse JavaScript.

*PostScript*
  Image classification.

---

### Thought Experiment: Image Recognition

How do you determine if an image is a landscape or a headshot?

Start with collecting data. You either need to get someone to label these things, which is a lot of work, or you can grab lots of pictures from flickr (*http://www.flickr.com/*) and ask for photos that have already been tagged.

Represent each image with a binned RGB (red, green, blue) intensity histogram. In other words, for each pixel, and for each of red, green, and blue, which are the basic colors in pixels, you measure the intensity, which is a number between 0 and 255.

Then draw three histograms, one for each basic color, showing how many pixels had which intensity. It's better to do a binned histogram, so have counts of the number of pixels of intensity 0-51, etc. In the end, for each picture, you have 15 numbers, corresponding to 3 colors and 5 bins per color. We are assuming here that every picture has the same number of pixels.

Finally, use k-NN to decide how much "blue" makes a landscape versus a headshot. You can tune the hyperparameters, which in this case are the number of bins as well as k.

---

# Jake's Exercise: Naive Bayes for Article Classification

This problem looks at an application of Naive Bayes for multiclass text classification. First, you will use the *New York Times* Developer API to fetch recent articles from several sections of the *Times*. Then, using

---

the simple Bernoulli model for word presence, you will implement a classifier which, given the text of an article from the *New York Times*, predicts the section to which the article belongs.

First, register for a *New York Times* Developer API key and request access to the Article Search API. After reviewing the API documentation, write code to download the 2,000 most recent articles for each of the Arts, Business, Obituaries, Sports, and World sections. (Hint: Use the nytd_section_facet facet to specify article sections.) The developer console may be useful for quickly exploring the API. Your code should save articles from each section to a separate file in a tab-delimited format, where the first column is the article URL, the second is the article title, and the third is the body returned by the API.

Next, implement code to train a simple Bernoulli Naive Bayes model using these articles. You can consider documents to belong to one of C categories, where the label of the $i$th document is encoded as $y_i \in 0, 1, 2, \ldots C$—for example, Arts = 0, Business = 1, etc.—and documents are represented by the sparse binary matrix $X$, where $X_{ij} = 1$ indicates that the $i$th document contains the $j$th word in our dictionary.

You train by counting words and documents within classes to estimate $\theta_{jc}$ and $\theta_c$:

$$\hat{\theta}_{jc} = \frac{n_{jc} + \alpha - 1}{n_c + \alpha + \beta - 2}$$

$$\hat{\theta}_c = \frac{n_c}{n}$$

where $n_{jc}$ is the number of documents of class $c$ containing the $j$th word, $n_c$ is the number of documents of class $c$, $n$ is the total number of documents, and the user-selected hyperparameters $\alpha$ and $\beta$ are pseudocounts that "smooth" the parameter estimates. Given these estimates and the words in a document $x$, you calculate the log-odds for each class (relative to the base class $c = 0$) by simply adding the class-specific weights of the words that appear to the corresponding bias term:

$$\log\left(\frac{p(y=c|x)}{p(y=0|x)}\right) = \sum_j \hat{w}_{jc} x_j + \hat{w}_{0c}$$

where

$$\hat{w}_{jc} = \log \frac{\hat{\theta}_{jc}\left(1 - \hat{\theta}_{j0}\right)}{\hat{\theta}_{j0}\left(1 - \hat{\theta}_{jc}\right)}$$

$$\hat{w}_{0c} = \sum_{j} \log \frac{1 - \hat{\theta}_{jc}}{1 - \hat{\theta}_{j0}} + \log \frac{\hat{\theta}_c}{\hat{\theta}_0}$$

Your code should read the title and body text for each article, remove unwanted characters (e.g., punctuation), and tokenize the article contents into words, filtering out stop words (given in the stopwords file). The training phase of your code should use these parsed document features to estimate the weights $\hat{w}$, taking the hyperparameters $\alpha$ and $\beta$ as input. The prediction phase should then accept these weights as inputs, along with the features for new examples, and output posterior probabilities for each class.

Evaluate performance on a randomized 50/50 train/test split of the data, including accuracy and runtime. Comment on the effects of changing $\alpha$ and $\beta$. Present your results in a (5×5) confusion table showing counts for the actual and predicted sections, where each document is assigned to its most probable section. For each section, report the top 10 most informative words. Also present and comment on the top 10 "most difficult to classify" articles in the test set.

Briefly discuss how you expect the learned classifier to generalize to other contexts, e.g., articles from other sources or time periods.

## Sample R Code for Dealing with the NYT API

```
# author: Jared Lander
#
# hard coded call to API
theCall <- "http://api.nytimes.com/svc/search/v1/
article?format=json&query=nytd_section_facet:
[Sports]&fields=url,title,body&rank=newest&offset=0
&api-key=Your_Key_Here"

# we need the rjson, plyr, and RTextTools packages
require(plyr)
require(rjson)
require(RTextTools)
```

```r
## first let's look at an individual call
res1 <- fromJSON(file=theCall)
# how long is the result
length(res1$results)
# look at the first item
res1$results[[1]]
# the first item's title
res1$results[[1]]$title
# the first item converted to a data.frame, Viewed in the data
viewer
View(as.data.frame(res1$results[[1]]))

# convert the call results into a data.frame, should be 10
rows by 3 columns
resList1 <- ldply(res1$results, as.data.frame)
View(resList1)

## now let's build this for multiple calls
# build a string where we will substitute the section for the
first %s and offset for the second %s
theCall <- "http://api.nytimes.com/svc/search/v1/
article?format=json&query=nytd_section_facet:
[%s]&fields=url,title,body&rank=newest&offset=%s
&api-key=Your_Key_Here"
# create an empty list to hold 3 result sets
resultsSports <- vector("list", 3)
## loop through 0, 1 and 2 to call the API for each value
for(i in 0:2)
{
    # first build the query string replacing the first %s with
    # Sport and the second %s with the current value of i
    tempCall <- sprintf(theCall, "Sports", i)
    # make the query and get the json response
    tempJson <- fromJSON(file=tempCall)
    # convert the json into a 10x3 data.frame and
    # save it to the list
    resultsSports[[i + 1]] <- ldply(tempJson$results,
    as.data.frame)
}
# convert the list into a data.frame
resultsDFSports <- ldply(resultsSports)
# make a new column indicating this comes from Sports
resultsDFSports$Section <- "Sports"

## repeat that whole business for arts
## ideally you would do this in a more eloquent manner, but
this is just for illustration
resultsArts <- vector("list", 3)
for(i in 0:2)
{
```

```
    tempCall <- sprintf(theCall, "Arts", i)
    tempJson <- fromJSON(file=tempCall)
    resultsArts[[i + 1]] <- ldply(tempJson$results,
    as.data.frame)
}
resultsDFArts <- ldply(resultsArts)
resultsDFArts$Section <- "Arts"

# combine them both into one data.frame
resultBig <- rbind(resultsDFArts, resultsDFSports)
dim(resultBig)
View(resultBig)

## now time for tokenizing
# create the document-term matrix in english, removing numbers
and stop words and stemming words
doc_matrix <- create_matrix(resultBig$body, language="english",
removeNumbers=TRUE, removeStopwords=TRUE, stemWords=TRUE)
doc_matrix
View(as.matrix(doc_matrix))

# create a training and testing set
theOrder <- sample(60)
container <- create_container(matrix=doc_matrix,
labels=resultBig$Section, trainSize=theOrder[1:40],
testSize=theOrder[41:60], virgin=FALSE)
```

### Historical Context: Natural Language Processing

The example in this chapter where the raw data is text is just the tip of the iceberg of a whole field of research in computer science called natural language processing (NLP). The types of problems that can be solved with NLP include machine translation, where given text in one language, the algorithm can translate the text to another language; semantic analysis; part of speech tagging; and document classification (of which spam filtering is an example). Research in these areas dates back to the 1950s.

# CHAPTER 5

# Logistic Regression

The contributor for this chapter is Brian Dalessandro. Brian works at Media6Degrees as a VP of data science, and he's active in the research community. He's also served as cochair of the KDD competition. M6D (also known as Media 6 Degrees) is a startup in New York City in the online advertising space. Figure 5-1 shows Brian's data science profile —his y-axis is scaled from Clown to Rockstar.

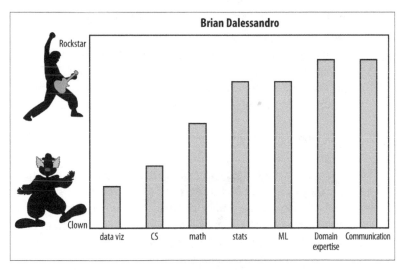

*Figure 5-1. Brian's data science profile*

Brian came to talk to the class about logistic regression and evaluation, but he started out with two thought experiments.

# Thought Experiments

1. How would data science differ if we had a "grand unified theory of everything"? Take this to mean a symbolic explanation of how the world works. This one question raises a bunch of other questions:

   - Would we even need data science if we had such a theory?

   - Is it even theoretically possible to have such a theory? Do such theories lie only in the realm of, say, physics, where we can anticipate the exact return of a comet we see once a century?

   - What's the critical difference between physics and data science that makes such a theory implausible?

   - Is it just accuracy? Or more generally, how much we imagine *can* be explained? Is it because we predict human behavior, which can be affected *by* our predictions, creating a feedback loop?

     It might be useful to think of the sciences as a continuum, where physics is all the way on the right, and as you go left, you get more chaotic—you're adding randomness (and salary). And where is economics on this spectrum? Marketing? Finance?

     If we could model this data science stuff like we already know how to model physics, we'd actually *know* when people will click on what ad, just as we know where the Mars Rover will land. That said, there's general consensus that the real world isn't as well-understood, nor do we expect it to be in the future.

2. In what sense does data science deserve the word "science" in its name?

   Never underestimate the power of creativity—people often have a vision they can describe but no method as to how to get there. As the data scientist, you have to turn that vision into a mathematical model within certain operational constraints. You need to state a well-defined problem, argue for a metric, and optimize for it as well. You also have to make sure you've actually answered the original question.

   There is *art* in data science—it's in translating human problems into the mathematical context of data science and back.

---

But we always have more than one way of doing this translation—more than one possible model, more than one associated metric, and possibly more than one optimization. So the *science* in data science is—given raw data, constraints, and a problem statement—how to navigate through that maze and make the best choices. Every design choice you make can be formulated as an hypothesis, against which you will use rigorous testing and experimentation to either validate or refute.

This process, whereby one formulates a well-defined hypothesis and then tests it, might rise to the level of a science in certain cases. Specifically, the scientific method is adopted in data science as follows:

- You hold on to your existing best performer.

- Once you have a new idea to prototype, set up an experiment wherein the two best models compete.

- Rinse and repeat (while not overfitting).

# Classifiers

This section focuses on the process of choosing a *classifier*. Classification involves mapping your data points into a finite set of labels or the probability of a given label or labels. We've already seen some examples of classification algorithms, such as Naive Bayes and k-nearest neighbors (k-NN), in the previous chapters. Table 5-1 shows a few examples of when you'd want to use classification:

*Table 5-1. Classifier example questions and answers*

| | |
|---|---|
| "Will someone click on this ad?" | 0 or 1 (no or yes) |
| "What number is this (image recognition)?" | 0, 1, 2, etc. |
| "What is this news article about?" | "Sports" |
| "Is this spam?" | 0 or 1 |
| "Is this pill good for headaches?" | 0 or 1 |

From now on we'll talk about binary classification only (0 or 1).

In this chapter, we're talking about logistic regression, but there's other classification algorithms available, including decision trees (which we'll cover in Chapter 7), random forests (Chapter 7), and support vector machines and neural networks (which we aren't covering in this book).

The big picture is that given data, a real-world classification problem, and constraints, you need to determine:

1. Which classifier to use
2. Which optimization method to employ
3. Which loss function to minimize
4. Which features to take from the data
5. Which evaluation metric to use

Let's talk about the first one: how do you know which classifier to choose? One possibility is to try them all, and choose the best performer. This is fine if you have no constraints, or if you ignore constraints. But usually constraints are a big deal—you might have tons of data, or not much time, or both. This is something people don't talk about enough. Let's look at some constraints that are common across most algorithms.

## Runtime

Say you need to update 500 models a day. That is the case at M6D, where their models end up being bidding decisions. In that case, they start to care about various speed issues. First, how long it takes to update a model, and second, how long it takes to use a model to actually make a decision if you have it. This second kind of consideration is usually more important and is called *runtime*.

Some algorithms are slow at runtime. For example, consider k-NN: given a new data point in some large-dimensional space, you actually have to find the k closest data points to it. In particular, you need to have all of your data points in memory.

Linear models, by contrast, are very fast, both to update and to use at runtime. As we'll see in Chapter 6, you can keep running estimates of the constituent parts and just update with new data, which is fast and doesn't require holding old data in memory. Once you have a linear model, it's just a matter of storing the coefficient vector in a runtime machine and doing a single dot product against the user's feature vector to get an answer.

# You

One underappreciated constraint of being a data scientist is your own understanding of the algorithm. Ask yourself carefully, do you understand it for real? *Really*? It's OK to admit it if you don't.

You don't have to be a master of every algorithm to be a good data scientist. The truth is, getting the best fit of an algorithm often requires intimate knowledge of said algorithm. Sometimes you need to tweak an algorithm to make it fit your data. A common mistake for people not completely familiar with an algorithm is to overfit when they think they're tweaking.

# Interpretability

You often need to be able to interpret your model for the sake of the business. Decision trees are very easy to interpret. Random forests, on the other hand, are not, even though they are almost the same thing. They can take exponentially longer to explain in full. If you don't have 15 years to spend understanding a result, you may be willing to give up some accuracy in order to have it be easier to understand.

For example, by law, credit card companies have to be able to explain their denial-of-credit decisions, so decision trees make more sense than random forests. You might not have a law about it where you work, but it still might make good sense for your business to have a simpler way to explain the model's decision.

# Scalability

How about scalability? In general, there are three things you have to keep in mind when considering scalability:

1. Learning time: How much time does it take to train the model?

2. Scoring time: How much time does it take to give a new user a score once the model is in production?

3. Model storage: How much memory does the production model use up?

Here's a useful paper to look at when comparing models: An Empirical Comparison of Supervised Learning Algorithms (*http://goo.gl/bLpoea*), from which we've learned:

- Simpler models are more interpretable but aren't as good performers.
- The question of which algorithm works best is problem-dependent.
- It's also constraint-dependent.

# M6D Logistic Regression Case Study

Brian and his team have three core problems as data scientists at M6D:

1. Feature engineering: Figuring out which features to use and how to use them.
2. User-level conversion prediction: Forecasting when someone will click.
3. Bidding: How much it is worth to show a given ad to a given user?

This case study focuses on the second problem. M6D uses logistic regression for this problem because it's highly scalable and works great for binary outcomes like clicks.

## Click Models

At M6D, they need to match clients, which represent advertising companies, to individual users. Generally speaking, the advertising companies want to target ads to users based on a user's likelihood to click. Let's discuss what kind of data they have available first, and then how you'd build a model using that data.

M6D keeps track of the websites users have visited, but the data scientists don't look at the contents of the page. Instead they take the associated URL and hash it into some random string. They thus accumulate information about users, which they stash in a vector. As an example, consider the user "u" during some chosen time period:

```
u = < &ltfxyz, 123, sdqwe, 13ms&gtg >
```

This means "u" visited four sites and the URLs that "u" visited are hashed to those strings. After collecting information like this for all the users, they build a giant matrix whose columns correspond to sites and whose rows correspond to users, and a given entry is "1" if that user went to that site. Note it's a sparse matrix, because most people don't go to all that many sites.

To make this a classification problem, they need to have *classes* they are trying to predict. Let's say in this case, they want to predict whether a given user will click on a shoe ad or not. So there's two classes: "users who clicked on the shoe ad" and "users who did not click on the shoe ad." In the training dataset, then, in addition to the sparse matrix described, they'll also have a variable, or column, of labels. They label the behavior "clicked on a shoe ad" as "1," say, and "didn't click" as "0." Once they fit the classifier to the dataset, for any new user, the classifier will predict whether he will click or not (the label) based on the predictors (the user's browsing history captured in the URL matrix).

Now it's your turn: your goal is to build and train the model from a training set. Recall that in Chapter 4 you learned about *spam classifiers*, where the features are words. But you didn't particularly care about the meaning of the words. They might as well be strings. Once you've labeled as described earlier, this looks just like spam classification because you have a binary outcome with a large sparse binary matrix capturing the predictors. You can now rely on well-established algorithms developed for spam detection.

You've reduced your current problem to a previously solved problem! In the previous chapter, we showed how to solve this with Naive Bayes, but here we'll focus on using logistic regression as the model.

The output of a logistic regression model is the *probability* of a given click in this context. Likewise, the spam filters really judge the *probability* of a given email being spam. You can use these probabilities directly or you could find a threshold so that if the probability is above that threshhold (say, 0.75), you predict a click (i.e., you show an ad), and below it you decide it's not worth it to show the ad. The point being here that unlike with linear regression—which does its best to predict the actual value—the aim of logistic regression isn't to predict the actual value (0 or 1). Its job is to output a probability.

Although technically it's possible to implement a linear model such as linear regression on such a dataset (i.e., R will let you do it and won't break or tell you that you shouldn't do it), one of the problems with a linear model like linear regression is that it would give predictions below 0 and above 1, so these aren't directly interpretable as probabilities.

## The Underlying Math

So far we've seen that the beauty of logistic regression is it outputs values bounded by 0 and 1; hence they can be directly interpreted as probabilities. Let's get into the math behind it a bit. You want a function that takes the data and transforms it into a single value bounded inside the closed interval $[0,1]$. For an example of a function bounded between 0 and 1, consider the inverse-logit function shown in Figure 5-2.

$$P(t) = \text{logit}^{-1}(t) \equiv \frac{1}{\left(1+e^{-t}\right)} = \frac{e^t}{1+e^t}$$

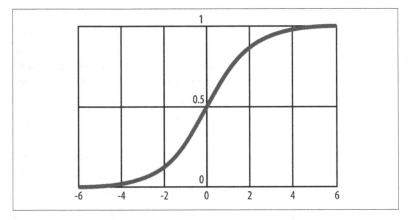

Figure 5-2. The inverse-logit function

---

## Logit Versus Inverse-logit

The logit function takes x values in the range $[0,1]$ and transforms them to y values along the entire real line:

$$logit(p) = log\left(\frac{p}{1-p}\right) = log(p) - log(1-p)$$

The inverse-logit does the reverse, and takes x values along the real line and tranforms them to y values in the range $[0,1]$.

---

Note when $t$ is large, $e^{-t}$ is tiny so the denominator is close to 1 and the overall value is close to 1. Similarly when $t$ is small, $e^{-t}$ is large so the denominator is large, which makes the function close to zero. So that's the inverse-logit function, which you'll use to begin deriving a logistic regression model. In order to model the data, you need to work with a slightly more general function that expresses the relationship between the data and a probability of a click. Start by defining:

$$P(c_i|x_i) = \left[\text{logit}^{-1}(\alpha + \beta^\tau x_i)\right]^{c_i} * \left[1 - \text{logit}^{-1}(\alpha + \beta^\tau x_i)\right]^{1-c_i}$$

Here $c_i$ is the labels or classes (clicked or not), and $x_i$ is the vector of features for user $i$. Observe that $c_i$ can only be 1 or 0, which means that if $c_i = 1$, the second term cancels out and you have:

$$P(c_i = 1|x_i) = \frac{1}{1 + e^{-(\alpha + \beta^\tau x_i)}} = \text{logit}^{-1}(\alpha + \beta^\tau x_i)$$

And similarly, if $c_i = 0$, the first term cancels out and you have:

$$P(c_i = 0|x_i) = 1 - \text{logit}^{-1}(\alpha + \beta^\tau x_i)$$

To make this a linear model in the outcomes $c_i$, take the log of the odds ratio:

$$\log(P(c_i = 1|x_i)/(1 - P(c_i = 1|x_i))) = \alpha + \beta^\tau x_i.$$

Which can also be written as:

$$\text{logit}(P(c_i = 1|x_i)) = \alpha + \beta^\tau x_i.$$

If it feels to you that we went in a bit of a circle here (this last equation was also implied by earlier equations), it's because we did. The purpose of this was to show you how to go back and forth between the probabilities and the linearity.

So the logit of the probability that user $i$ clicks on the shoe ad is being modeled as a linear function of the features, which were the URLs that user $i$ visited. This model is called the *logistic regression* model.

The parameter $\alpha$ is what we call the *base rate*, or the unconditional probability of "1" or "click" knowing nothing more about a given user's feature vector $x_i$. In the case of measuring the likelihood of an average user clicking on an ad, the base rate would correspond to the click-through rate, i.e., the tendency over all users to click on ads. This is typically on the order of 1%.

If you had no information about your specific situation except the base rate, the average prediction would be given by just $\alpha$:

$$P\left(c_i = 1\right) = \frac{1}{1 + e^{-\alpha}}$$

The variable $\beta$ defines the slope of the logit function. Note that in general it's a vector that is as long as the number of features you are using for each data point. The vector $\beta$ determines the extent to which certain features are markers for increased or decreased likelihood to click on an ad.

## Estimating $\alpha$ and $\beta$

Your immediate modeling goal is to use the training data to find the best choices for $\alpha$ and $\beta$. In general you want to solve this with maximum likelihood estimation and use a convex optimization algorithm because the likelihood function is convex; you can't just use derivatives and vector calculus like you did with linear regression because it's a complicated function of your data, and in particular there is no closed-form solution.

Denote by $\Theta$ the pair $\{\alpha, \beta\}$. The *likelihood function L* is defined by:

$$L\left(\Theta \mid X_1, X_2, \cdots, X_n\right) = P\left(X \mid \Theta\right) = P\left(X_1 \mid \Theta\right) \cdots\cdots P\left(X_n \mid \Theta\right)$$

where you are assuming the data points $X_i$ are independent, where $i = 1, \ldots, n$ represent your $n$ users. This independence assumption corresponds to saying that the click behavior of any given user doesn't affect the click behavior of all the other users—in this case, "click behavior" means "probability of clicking." It's a relatively safe assumption at a given point in time, but not forever. (Remember the independence assumption is what allows you to express the likelihood function as the product of the densities for each of the $n$ observations.)

You then search for the parameters that maximize the likelihood, having observed your data:

$$\Theta_{MLE} = argmax_{\Theta} \; \Pi_1^n \; P(X_i|\Theta).$$

Setting $p_i = 1 / \left( 1 + e^{-\left( \alpha + \beta^t x_i \right)} \right)$, the probability of a single observation, $P(X_i|\Theta)$ is:

$$p_i^{c_i} \cdot (1 - p_i)^{1-c_i}$$

So putting it all together, you have:

$$\Theta_{MLE} = argmax_{\Theta} \; \Pi_1^n \; p_i^{c_i} \cdot (1 - p_i)^{1-c_i}$$

Now, how do you maximize the likelihood?

Well, when you faced this situation with linear regression, you took the derivative of the likelihood with respect to $\alpha$ and $\beta$, set that equal to zero, and solved. But if you try that in this case, it's not possible to get a closed-form solution. The key is that the values that maximize the likelihood will also maximize the log likelihood, which is equivalent to minimizing the *negative* log likelihood. So you transform the problem to find the minimum of the negative log likelihood.

Now you have to decide which optimization method to use. As it turns out, under reasonable conditions, both of the optimization methods we describe below will converge to a *global maximum* when they converge at all. The "reasonable condition" is that the variables are not linearly dependent, and in particular guarantees that the Hessian matrix will be positive definite.

**More on Maximum Likelihood Estimation**
We realize that we went through that a bit fast, so if you want more details with respect to maximum likelihood estimation, we suggest looking in *Statistical Inference* by Casella and Berger, or if it's linear algebra in general you want more details on, check out Gilbert Strang's *Linear Algebra and Its Applications*.

## Newton's Method

You can use numerical techniques to find your global maximum by following the reasoning underlying Newton's method from calculus; namely, that you can pretty well approximate a function with the first few terms of its Taylor Series.

Specifically, given a step size $\gamma$, you compute a local gradient $\triangledown\Theta$, which corresponds to the first derivative, and a Hessian matrix $H$, which corresponds to the second derivative. You put them together and each step of the algorithm looks something like this:

$$\Theta_{n+1} = \Theta_n - \gamma H^{-1} \cdot \triangledown\Theta.$$

Newton's method uses the curvature of log-likelihood to choose an appropriate step direction. Note that this calculation involves inverting the $k \times k$ Hessian matrix, which is bad when there's lots of features, as in 10,000 or something. Typically you don't have that many features, but it's not impossible.

In practice you'd never actually invert the Hessian—instead you'd solve an equation of the form $Ax = y$, which is much more computationally stable than finding $A^{-1}$.

## Stochastic Gradient Descent

Another possible method to maximize your likelihood (or minimize your negative log likelihood) is called Stochastic Gradient Descent. It approximates a gradient using a single observation at a time. The algorithm updates the current best-fit parameters each time it sees a new data point. The good news is that there's no big matrix inversion, and it works well with both huge data and sparse features; it's a big deal in Mahout (*http://mahout.apache.org*) and Vowpal Wabbit (*http://hunch.net/~vw/*), two open source projects to enable large-scale machine learning across a variety of algorithms. The bad news is it's not such a great optimizer and it's very dependent on step size.

## Implementation

In practice you don't have to code up the iterative reweighted least squares or stochastic gradient optimization method yourself; this is implemented in R or in any package that implements logistic regression. So suppose you had a dataset that had the following first five rows:

```
click url_1 url_2 url_3 url_4 url_5
1     0     0     0     1     0
1     0     1     1     0     1
0     1     0     0     1     0
1     0     0     0     0     0
1     1     0     1     0     1
```

Call this matrix "train," and then the command line in R would be:

```
fit <- glm(click ~ url_1 + url_2 + url_3 + url_4 + url_5,
           data = train, family = binomial(logit))
```

# Evaluation

Let's go back to the big picture from earlier in the chapter where we told you that you have many choices you need to make when confronted with a classification problem. One of the choices is how you're going to evaluate your model. We discussed this already in Chapter 3 with respect to linear regression and k-NN, as well as in the previous chapter with respect to Naive Bayes. We generally use different evaluation metrics for different kinds of models, and in different contexts. Even logistic regression can be applied in multiple contexts, and depending on the context, you may want to evaluate it in different ways.

First, consider the context of using logistic regression as a ranking model—meaning you are trying to determine the *order* in which you show ads or items to a user based on the probability they would click. You could use logistic regression to estimate probabilities, and then rank-order the ads or items in decreasing order of likelihood to click based on your model. If you wanted to know how good your model was at discovering relative rank (notice in this case, you could care less about the absolute scores), you'd look to one of:

*Area under the receiver operating curve (AUC)*
    In signal detection theory, a receiver operating characteristic curve, or ROC curve, is defined as a plot of the true positive rate against the false positive rate for a binary classification problem as you change a threshold. In particular, if you took your training set and ranked the items according to their probabilities and varied the threshold (from $\infty$ to $-\infty$) that determined whether to classify the item as 1 or 0 , and kept plotting the true positive rate versus the false positive rate, you'd get the ROC curve. The area under that curve, referred to as the AUC, is a way to measure the

success of a classifier or to compare two classifiers. Here's a nice paper on it by Tom Fawcett, "Introduction to ROC Analysis" (*http://goo.gl/fE8i7s*).

*Area under the cumulative lift curve*
An alternative is the area under the cumulative lift curve, which is frequently used in direct marketing and captures how many times it is better to use a model versus not using the model (i.e., just selecting at random).

We'll see more of both of these in Chapter 13.

---

## Warning: Feedback Loop!

If you want to productionize logistic regression to rank ads or items based on clicks and impressions, then let's think about what that means in terms of the data getting generated. So let's say you put an ad for hair gel above an ad for deodorant, and then more people click on the ad for hair gel—is it because you put it on top or because more people want hair gel? How can you feed that data into future iterations of your algorithm given that you potentially caused the clicks yourself and it has nothing to do with the ad quality? One solution is to always be logging the *position* or *rank* that you showed the ads, and then use *that* as one of the predictors in your algorithm. So you would then model the probability of a click as a function of position, vertical, brand, or whatever other features you want. You could then use the parameter estimated for position and use that going forward as a "position normalizer." There's a whole division at Google called Ads Quality devoted to problems such as these, so one little paragraph can't do justice to all the nuances of it.

---

Second, now suppose instead that you're using logistic regression for the purposes of classification. Remember that although the observed output was a binary outcome (1,0), the output of a logistic model is a probability. In order to use this for classification purposes, for any given unlabeled item, you would get its predicted probability of a click. Then to minimize the misclassification rate, if the predicted probability is > 0.5 that the label is 1 (click), you would label the item a 1 (click), and otherwise 0. You have several options for how you'd then evaluate the quality of the model, some of which we already discussed in Chapters 3 and 4, but we'll tell you about them again here so you see their pervasiveness:

*Lift*

How much more people are buying or clicking because of a model (once we've introduced it into production).

*Accuracy*

How often the correct outcome is being predicted, as discussed in Chapters 3 and 4.

*Precision*

This is the (number of true positives)/(number of true positives + number of false positives).

*Recall*

This is the (number of true positives)/(number of true positives + number of false negatives).

*F-score*

We didn't tell you about this one yet. It essentially combines precision and recall into a single score. It's the harmonic mean of precision and recall, so $(2 \times precision \times recall)/(precision + recall)$. There are generalizations of this that are essentially changing how much you weight one or the other.

Finally, for density estimation, where we need to know an actual probability rather than a relative score, we'd look to:

*Mean squared error*

We discussed with respect to linear regression. As a reminder, this is the average squared distance between the predicted and actual values.

*Root squared error*

The square root of mean squared error.

*Mean absolute error*

This is a variation on mean squared error and is simply the average of the absolute value of the difference between the predicted and actual values.

In general, it's hard to compare lift curves, but you can compare AUC (area under the receiver operator curve)—they are "base rate invariant." In other words, if you bring the click-through rate from 1% to 2%, that's 100% lift; but if you bring it from 4% to 7%, that's less lift but more effect. AUC does a better job in such a situation when you want to compare.

Density estimation tests tell you how well are you fitting for conditional probability. In advertising, this may arise if you have a situation where each ad impression costs $c and for each conversion you receive $q. You will want to target every conversion that has a positive expected value, i.e., whenever:

$$P(Conversion|X) \cdot \$q > \$c$$

But to do this you need to make sure the probability estimate on the left is *accurate*, which in this case means something like the mean squared error of the estimator is small. Note a model can give you good relative rankings—it gets the order right—but bad estimates of the probability.

Think about it this way: it could say to rank items in order 1, 2, and 3 and estimate the probabilities as .7, .5, and .3. It might be that the *true* probabilities are .03, .02, and .01, so our estimates are totally off, but the ranking was correct.

---

### Using A/B Testing for Evaluation

When we build models and optimize them with respect to some evaluation metric such as accuracy or mean squared error, the estimation method itself is built to optimize parameters *with respect* to these metrics. In some contexts, the metrics we might want to optimize for are something else altogether, such as revenue. So we might try to build an algorithm that optimizes for accuracy, when our real goal is to make money. The model itself may not directly capture this. So a way to capture it is to run A/B tests (or statistical experiments) where we divert some set of users to one version of the algorithm and another set of users to another version of the algorithm, and check the difference in performance of metrics we care about, such as revenue or revenue per user, or something like that. We'll discuss A/B testing more in Chapter 11.

---

## Media 6 Degrees Exercise

Media 6 Degrees kindly provided a dataset that is perfect for exploring logistic regression models, and evaluating how good the models are. Follow along by implementing the following R code. The dataset can be found at *https://github.com/oreillymedia/doing_data_science*.

---

# Sample R Code

```r
# Author: Brian Dalessandro
# Read in data, look at the variables and create a training
and test set
file <- "binary_class_dataset.txt"
set <- read.table(file, header = TRUE, sep = "\t",
                  row.names = "client_id")
names(set)

split <- .65
set["rand"] <- runif(nrow(set))
train <- set[(set$rand <= split), ]
test <- set[(set$rand > split), ]
set$Y <- set$Y_BUY


#############################################################
##########            R FUNCTIONS            ##########
#############################################################

library(mgcv)

# GAM Smoothed plot
plotrel <- function(x, y, b, title) {
    # Produce a GAM smoothed representation of the data
    g <- gam(as.formula("y ~ x"), family = "binomial",
            data = set)
    xs <- seq(min(x), max(x), length = 200)
    p <- predict(g, newdata = data.frame(x = xs),
                 type = "response")

    # Now get empirical estimates (and discretize if
    # non discrete)
    if (length(unique(x)) > b) {
        div <- floor(max(x) / b)
        x_b <- floor(x / div) * div
        c <- table(x_b, y)
    }
    else { c <- table(x, y) }
    pact <- c[ , 2]/(c[ , 1]+c[ , 2])
    cnt <- c[ , 1]+c[ , 2]
    xd <- as.integer(rownames(c))
    plot(xs, p, type="l", main=title,
        ylab = "P(Conversion | Ad, X)", xlab="X")
        points(xd, pact, type="p", col="red")
    rug(x+runif(length(x)))
}


library(plyr)
```

```
# wMAE plot and calculation
getmae <- function(p, y, b, title, doplot) {
    # Normalize to interval [0,1]
    max_p <- max(p)
    p_norm <- p / max_p
    # break up to b bins and rescale
    bin <- max_p * floor(p_norm * b) / b
    d <- data.frame(bin, p, y)
    t <- table(bin)
    summ <- ddply(d, .(bin), summarise, mean_p = mean(p),
                  mean_y = mean(y))
    fin <- data.frame(bin = summ$bin, mean_p = summ$mean_p,
                      mean_y = summ$mean_y, t)
    # Get wMAE
    num = 0
    den = 0
    for (i in c(1:nrow(fin))) {
        num <- num + fin$Freq[i] * abs(fin$mean_p[i] -
                                       fin$mean_y[i])
        den <- den + fin$Freq[i]
    }
    wmae <- num / den
    if (doplot == 1) {
        plot(summ$bin, summ$mean_p, type = "p",
            main = paste(title," MAE =", wmae),
            col = "blue", ylab = "P(C | AD, X)",
            xlab = "P(C | AD, X)")
        points(summ$bin, summ$mean_y, type = "p", col = "red")
        rug(p)
    }
    return(wmae)
}


library(ROCR)
get_auc <- function(ind, y) {
    pred <- prediction(ind, y)
    perf <- performance(pred, 'auc', fpr.stop = 1)
    auc <- as.numeric(substr(slot(perf, "y.values"), 1, 8),
                      double)
    return(auc)
}


# Get X-Validated performance metrics for a given feature set

getxval <- function(vars, data, folds, mae_bins) {
    # assign each observation to a fold
    data["fold"] <- floor(runif(nrow(data)) * folds) + 1
    auc <- c()
    wmae <- c()
```

```r
    fold <- c()
    # make a formula object
    f = as.formula(paste("Y", "~", paste(vars,
                  collapse = "+")))
    for (i in c(1:folds)) {
        train <- data[(data$fold != i), ]
        test <- data[(data$fold == i), ]
        mod_x <- glm(f, data=train, family = binomial(logit))
        p <- predict(mod_x, newdata = test, type = "response")
        # Get wMAE
        wmae <- c(wmae, getmae(p, test$Y, mae_bins,
                  "dummy", 0))
        fold <- c(fold, i)
        auc <- c(auc, get_auc(p, test$Y))
    }
    return(data.frame(fold, wmae, auc))
}


##################################################################
##########          MAIN: MODELS AND PLOTS          ##########
##################################################################
# Now build a model on all variables and look at coefficients
and model fit
vlist      <-      c("AT_BUY_BOOLEAN",      "AT_FREQ_BUY",
"AT_FREQ_LAST24_BUY",
    "AT_FREQ_LAST24_SV", "AT_FREQ_SV", "EXPECTED_TIME_BUY",
    "EXPECTED_TIME_SV", "LAST_BUY", "LAST_SV", "num_checkins")
f = as.formula(paste("Y_BUY", "~" , paste(vlist,
            collapse = "+")))
fit <- glm(f, data = train, family = binomial(logit))
summary(fit)

# Get performance metrics on each variable

vlist      <-      c("AT_BUY_BOOLEAN",      "AT_FREQ_BUY",
"AT_FREQ_LAST24_BUY",
    "AT_FREQ_LAST24_SV", "AT_FREQ_SV", "EXPECTED_TIME_BUY",
    "EXPECTED_TIME_SV", "LAST_BUY", "LAST_SV", "num_checkins")

# Create empty vectors to store the performance/evaluation met
rics
auc_mu <- c()
auc_sig <- c()
mae_mu <- c()
mae_sig <- c()

for (i in c(1:length(vlist))) {
    a <- getxval(c(vlist[i]), set, 10, 100)
    auc_mu <- c(auc_mu, mean(a$auc))
    auc_sig <- c(auc_sig, sd(a$auc))
    mae_mu <- c(mae_mu, mean(a$wmae))
```

```
    mae_sig <- c(mae_sig, sd(a$wmae))
}

univar <- data.frame(vlist, auc_mu, auc_sig, mae_mu, mae_sig)

# Get MAE plot on single variable -
# use holdout group for evaluation
set <- read.table(file, header = TRUE, sep = "\t",
                  row.names="client_id")
names(set)

split<-.65
set["rand"] <- runif(nrow(set))
train <- set[(set$rand <= split), ]
test <- set[(set$rand > split), ]
set$Y <- set$Y_BUY

fit <- glm(Y_BUY ~ num_checkins, data = train,
           family = binomial(logit))
y <- test$Y_BUY
p <- predict(fit, newdata = test, type = "response")

getmae(p,y,50,"num_checkins",1)

# Greedy Forward Selection
rvars <- c("LAST_SV", "AT_FREQ_SV", "AT_FREQ_BUY",
    "AT_BUY_BOOLEAN", "LAST_BUY", "AT_FREQ_LAST24_SV",
    "EXPECTED_TIME_SV", "num_checkins",
    "EXPECTED_TIME_BUY", "AT_FREQ_LAST24_BUY")
# Create empty vectors
auc_mu <- c()
auc_sig <- c()
mae_mu <- c()
mae_sig <- c()

for (i in c(1:length(rvars))) {
    vars <- rvars[1:i]
    vars
    a <- getxval(vars, set, 10, 100)
    auc_mu <- c(auc_mu, mean(a$auc))
    auc_sig <- c(auc_sig, sd(a$auc))
    mae_mu <- c(mae_mu, mean(a$wmae))
    mae_sig <- c(mae_sig, sd(a$wmae))
}
kvar<-data.frame(auc_mu, auc_sig, mae_mu, mae_sig)

# Plot 3 AUC Curves
y <- test$Y_BUY

fit <- glm(Y_BUY~LAST_SV, data=train,
           family = binomial(logit))
```

```
p1 <- predict(fit, newdata=test, type="response")
fit <- glm(Y_BUY~LAST_BUY, data=train,
           family = binomial(logit))
p2 <- predict(fit, newdata=test, type="response")
fit <- glm(Y_BUY~num_checkins, data=train,
           family = binomial(logit))
p3 <- predict(fit, newdata=test,type="response")

pred <- prediction(p1,y)
perf1 <- performance(pred,'tpr','fpr')
pred <- prediction(p2,y)
perf2 <- performance(pred,'tpr','fpr')
pred <- prediction(p3,y)
perf3 <- performance(pred,'tpr','fpr')


plot(perf1, color="blue", main="LAST_SV (blue),
     LAST_BUY (red), num_checkins (green)")
plot(perf2, col="red", add=TRUE)
plot(perf3, col="green", add=TRUE)
```

CHAPTER 6
# Time Stamps and Financial Modeling

In this chapter, we have two contributors, Kyle Teague from GetGlue, and someone you are a bit more familiar with by now: Cathy O'Neil. Before Cathy dives into her talk about the main topics for this chapter —times series, financial modeling, and fancypants regression—we'll hear from Kyle Teague from GetGlue about how they think about building a recommendation system. (We'll also hear more on this topic in Chapter 7.) We then lay some of the groundwork for thinking about timestamped data, which will segue into Cathy's talk.

## Kyle Teague and GetGlue

We got to hear from Kyle Teague, a VP of data science and engineering at GetGlue. Kyle's background is in electrical engineering. He considers the time he spent doing signal processing in research labs as super valuable, and he's been programming since he was a kid. He develops in Python.

GetGlue is a New York-based startup whose primary goal is to address the problem of content discovery within the movie and TV space. The usual model for finding out what's on TV is the 1950's *TV Guide* schedule; that's still how many of us find things to watch. Given that there are thousands of channels, it's getting increasingly difficult to find out what's good on TV.

GetGlue wants to change this model, by giving people personalized TV recommendations and personalized guides. Specifically, users "check in" to TV shows, which means they can tell other people they're watching a show, thereby creating a timestamped data point. They can also perform other actions such as liking or commenting on the show.

We store information in triplets of data of the form {user, action, item}, where the item is a TV show (or a movie). One way to visualize this stored data is by drawing a *bipartite graph* as shown in Figure 6-1.

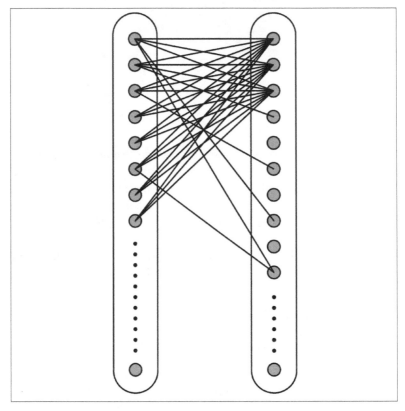

*Figure 6-1. Bipartite graph with users and items (shows) as nodes*

We'll go into graphs in later chapters, but for now you should know that the dots are called "nodes" and the lines are called "edges." This specific kind of graph, called a *bipartite graph*, is characterized by there being two kinds of nodes, in this case corresponding to "users" and "items." All the edges go between a user and an item, specifically if the user in question has acted in some way on the show in question. There are never edges between different users or different shows. The graph in Figure 6-1 might display when certain users have "liked" certain TV shows.

GetGlue enhances the graph as follows: it finds ways to create edges between users and between shows, albeit with different kinds of edges. So, for example, users can follow one another or be friends on GetGlue, which produces *directed edges*; i.e., edges going from one node to another with a direction, usually denoted by an arrow. Similarly, using the set of preferences, GetGlue can learn that two people have similar tastes and they can be connected that way as well, which would probably not be directed.

GetGlue also hires human evaluators to make connections or directional edges between shows. So, for example, *True Blood* and *Buffy the Vampire Slayer* might be similar for some reason, and so the humans create a "similarity" edge in the graph between them. There are nuances around the edge being directional; they may draw an arrow pointing from *Buffy* to *True Blood* but not vice versa, for example, so their notion of "similar" or "close" captures both content and popularity. Pandora is purported to do something like this, too.

Another important aspect, especially in the fast-paced TV space, is time. The user checked in or liked a show at a specific time, so the data that logs an action needs to have a timestamp as well: {user, action, item, timestamp}. Timestamps are helpful to see how influence propagates in the graph we've constructed, or how the graph evolves over time.

# Timestamps

Timestamped event data is a common data type in the age of Big Data. In fact, it's one of the causes of Big Data. The fact that computers can record all the actions a user takes means that a single user can generate thousands of data points alone in a day. When people visit a website or use an app, or interact with computers and phones, their actions can be logged, and the exact time and nature of their interaction recorded. When a new product or feature is built, engineers working on it write code to capture the events that occur as people navigate and use the product—that capturing is part of the product.

For example, imagine a user visits the *New York Times* home page. The website captures which news stories are rendered for that user, and which stories were clicked on. This generates event logs. Each record is an event that took place between a user and the app or website.

Here's an example of raw data point from GetGlue:

```
{"userId": "rachelschutt", "numCheckins": "1",
"modelName": "movies", "title": "Collaborator",
"source": "http://getglue.com/stickers/tribeca_film/
collaborator_coming_soon", "numReplies": "0",
"app": "GetGlue", "lastCheckin": "true",
"timestamp": "2012-05-18T14:15:40Z",
"director": "martin donovan", "verb": "watching",
"key": "rachelschutt/2012-05-18T14:15:40Z",
"others": "97", "displayName": "Rachel Schutt",
"lastModified": "2012-05-18T14:15:43Z",
"objectKey": "movies/collaborator/martin_donovan",
"action": "watching"}
```

If we extract four fields: {"userid":"rachelschutt", "action": "watching", "title":"Collaborator", timestamp:"2012-05-18T14:15:40Z" }, we can think of it as being in the order we just discussed, namely {user, verb, object, timestamp}.

## Exploratory Data Analysis (EDA)

As we described in Chapter 2, it's best to start your analysis with EDA so you can gain intuition for the data before building models with it. Let's delve deep into an example of EDA you can do with user data, stream-of-consciousness style. This is an illustration of a larger technique, and things we do here can be modified to other types of data, but you also might need to do something else entirely depending on circumstances.

The very first thing you should look into when dealing with user data is *individual user plots over time*. Make sure the data makes sense to you by investigating the narrative the data indicates from the perspective of *one person*.

To do this, take a random sample of users: start with something small like 100 users. Yes, maybe your dataset has millions of users, but to start out, you need to gain intuition. Looking at millions of data points is too much for you as a human. But just by looking at 100, you'll start to understand the data, and see if it's clean. Of course, this kind of sample size is not large enough if you were to start making inferences about the entire set of data.

You might do this by finding usernames and grepping or searching for 100 random choices, one at a time. For each user, create a plot like the one in Figure 6-2.

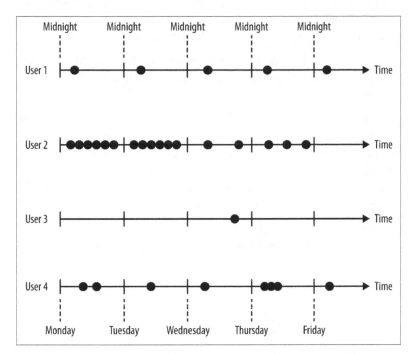

*Figure 6-2. An example of a way to visually display user-level data over time*

Now try to construct a narrative from that plot. For example, we could say that user 1 comes the same time each day, whereas user 2 started out active in this time period but then came less and less frequently. User 3 needs a longer time horizon for us to understand his or her behavior, whereas user 4 looks "normal," whatever that means.

Let's pose questions from that narrative:

- What *is* the typical or average user doing?
- What does variation around that look like?
- How would we classify users into different segments based on their behavior with respect to time?
- How would we quantify the differences between these users?

Think abstractly about a certain typical question from data munging discipline. Say we have some raw data where each data point is an event, but we want to have data stored in rows where each row consists of a user followed by a bunch of timestamps corresponding to actions that user performed. How would we get the data to that point? Note that different users will have a different number of timestamps.

Make this reasoning explicit: how would we write the code to create a plot like the one just shown? How would we go about tackling the data munging exercise?

Suppose a user can take multiple actions: "thumbs_up," or "thumbs_down," "like," and "comment." How can we plot those events? How can we modify our metrics? How can we encode the user data with these different actions? Figure 6-3 provides an example for the first question where we color code actions thumbs up and thumbs down, denoted thumbs_up and thumbs_down.

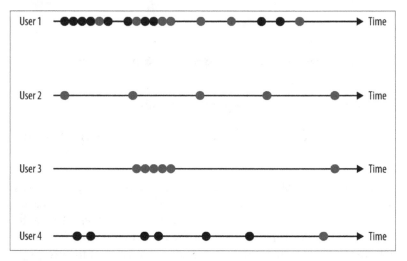

*Figure 6-3. Use color to include more information about user actions in a visual display*

In this toy example, we see that all the users did the same thing at the same time toward the right end of the plots. Wait, is this a real event or a bug in the system? How do we check that? Is there a large co-occurence of some action across users? Is "black" more common than "red"? Maybe some users like to always thumb things up, another group always like to thumb things down, and some third group of users are a mix. What's the definition of a "mix"?

Now that we've started to get some sense of variation across users, we can think about how we might want to *aggregate* users. We might make the x-axis refer to time, and the y-axis refer to counts, as shown in Figure 6-4.

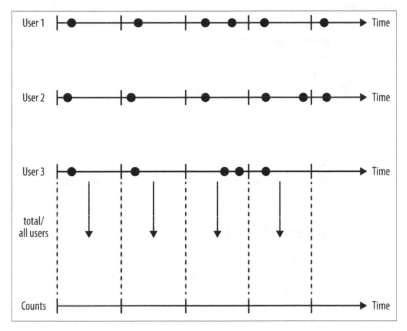

*Figure 6-4. Aggregating user actions into counts*

We're no longer working with 100 individual users, but we're still making choices, and those choices will impact our perception and understanding of the dataset.

For example, are we counting the number of unique users or the overall number of user logins? Because some users log in multiple times, this can have a huge impact. Are we counting the number of actions or the number of users who did a given action at least once during the given time segment?

What is our time horizon? Are we counting per second, minute, hour, 8-hour segments, day, or week? Why did we choose that? Is the signal overwhelmed by seasonality, or are we searching *for* seasonality?

Are our users in different time zones? If user 1 and user 2 are in New York and London, respectively, then it's 7 a.m. in NYC when it's noon in London. If we count that as 7 a.m. and say "30,000 users did this action in the morning," then it's misleading, because it's not morning in London. How are we going to treat this? We could shift the data into buckets so it's 7 a.m. for the *user*, not 7 a.m. in New York, but then we encounter other problems. This is a decision we have to make and justify, and it could go either way depending on circumstances.

**Timestamps Are Tricky**

We're not gonna lie, timestamps are one of the hardest things to get right about modeling, especially around time changes. That's why it's sometimes easiest to convert all timestamps into seconds since the beginning of epoch time (*http://www.epoch converter.com/*).

Maybe we want to make different plots for different action types, or maybe we will bin actions into broader categories to take a good look at them and gain perspective.

## Metrics and New Variables or Features

The intuition we gained from EDA can now help us construct metrics. For example, we can measure users by keeping tabs on the frequencies or counts of their actions, the time to first event, or simple binary variables such as "did action at least once," and so on. If we want to compare users, we can construct similarity or difference metrics. We might want to aggregate by user by counting or having a cumulative sum of counts or money spent, or we might aggregate by action by having cumulative or average number of users doing that action once or more than once.

This is by no means a comprehensive list. Metrics can be any function of the data, as long as we have a purpose and reason for creating them and interpreting them.

## What's Next?

We want to start moving toward modeling, algorithms, and analysis, incorporating the intuition we built from the EDA into our models and algorithms. Our next steps depend on the context, but here are some examples of what we could do.

We might be interested in time series modeling, which includes auto-regression. We'll talk about this more in the next section on financial modeling, but generally speaking we work with time series when we are trying to predict events that are super time-sensitive, like markets, or somewhat predictable based on what already happened, such as how much money gets invested in pension funds per month.

We might start clustering, as we discussed in Chapter 3. In order to do this, we'd need to define the *closeness* of users with each other.

Maybe we'll want to build a monitor that could automatically detect common behavior patterns. Of course we'd first have to define what a common behavior pattern is and what would make things uncommon.

We might try our hands at change-point detection, which is to say the ability to identify when some big event has taken place. What kind of behavior in our system should trigger an alarm? Or, we might try to establish causality. This can be very hard if we haven't set up an experiment. Finally, we might want to train a recommendation system.

---

## Historical Perspective: What's New About This?

Timestamped data itself is not new, and time series analysis is a well-established field (see, for example, *Time Series Analysis* by James D. Hamilton). Historically, the available datasets were fairly small and events were recorded once a day, or even reported at aggregate levels. Some examples of timestamped datasets that have existed for a while even at a granular level are stock prices in finance, credit card transactions, phone call records, or books checked out of the library.

Even so, there are a couple things that make this new, or at least the scale of it new. First, it's now easy to measure human behavior throughout the day because many of us now carry around devices that can be and are used for measurement purposes and to record actions. Next, timestamps are accurate, so we're not relying on the user to self-report, which is famously unreliable. Finally, computing power makes it possible to store large amounts of data and process it fairly quickly.

---

# Cathy O'Neil

Next Cathy O'Neil spoke. You're already familiar with her, but let's check out her data science profile in Figure 6-5.

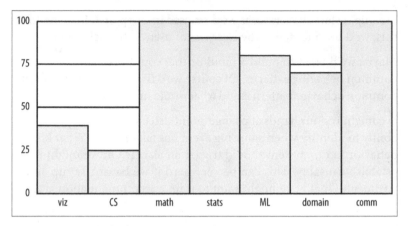

*Figure 6-5. Cathy's data science profile*

Her most obvious weakness is in CS. Although she programs in Python pretty proficiently, and can scrape and parse data, prototype models, and use matplotlib to draw pretty pictures, she is no Java map-reducer and bows down to those people who are. She's also completely untrained in data visualization, but knows enough to get by and give presentations that people understand.

# Thought Experiment

*What do you lose when you think of your training set as a big pile of data and ignore the timestamps?*

The big point here is that you can't tease out cause and effect if you don't have any sense of time.

What if we amend the question to allow the collection of *relative time differentials*, so "time since user last logged in" or "time since last click" or "time since last insulin injection," but not *absolute timestamps*?

In that case you still have major problems. For example, you'd ignore trends altogether, as well as seasonality, if you don't order your data over time. So for the insulin example, you might note that 15 minutes after your insulin injection your blood sugar goes down consistently, but you might *not* notice an overall trend of your rising blood sugar over the past few months if your dataset for the past few months has no absolute timestamp on it. Without putting this data in order, you'd miss the pattern shown in Figure 6-6.

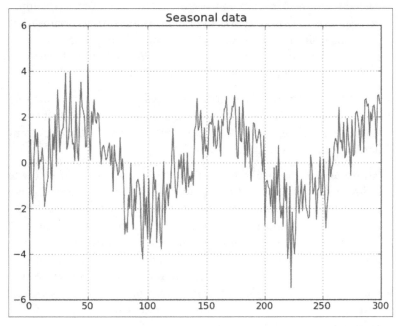

*Figure 6-6. Without keeping track of timestamps, we can't see time-based patterns; here, we see a seasonal pattern in a time series*

This idea, of keeping track of trends and seasonalities, is very important in financial data, and essential to keep track of if you want to make money, considering how small the signals are.

# Financial Modeling

Before the term data scientist existed, there were quants working in finance. There are many overlapping aspects of the job of the quant and the job of the data scientist, and of course some that are very different. For example, as we will see in this chapter, quants are singularly obsessed with timestamps, and don't care much about *why* things work, just if they do.

Of course there's a limit to what can be covered in just one chapter, but this is meant to give a taste of the kind of approach common in financial modeling.

## In-Sample, Out-of-Sample, and Causality

We need to establish a strict concept of in-sample and out-of-sample data. Note the out-of-sample data is *not* meant as testing data—that all happens inside in-sample data. Rather, out-of-sample data is meant to be the data you use after *finalizing your model* so that you have some idea how the model will perform in production.

We should even restrict the number of times one does out-of-sample analysis on a given dataset because, like it or not, we learn stuff about that data every time, and we will subconsciously overfit to it even in different contexts, with different models.

Next, we need to be careful to always perform *causal modeling* (note this differs from what statisticians mean by causality). Namely, *never use information in the future to predict something now*. Or, put differently, we only use information from the past up and to the present moment to predict the future. This is incredibly important in financial modeling. Note it's *not enough* to use data *about* the present if it isn't actually available and accessible at the present moment. So this means we have to be very careful with timestamps of availability as well as timestamps of reference. This is huge when we're talking about lagged government data.

Similarly, when we have a set of training data, we don't know the "best-fit coefficients" for that training data until after the last timestamp on all the data. As we move forward in time from the first timestamp to the last, we expect to get different sets of coefficients as more events happen.

One consequence of this is that, instead of getting *one* set of "best-fit" coefficients, we actually get an *evolution* of each coefficient. This is helpful because it gives us a sense of how stable those coefficients are. In particular, if one coefficient has changed sign 10 times over the training set, then we might well expect a good estimate for it is zero, not the so-called "best fit" at the end of the data. Of course, depending on the variable, we might think of a legitimate reason for it to actually change sign over time.

The in-sample data should, generally speaking, come before the out-of-sample data to avoid causality problems as shown in Figure 6-7.

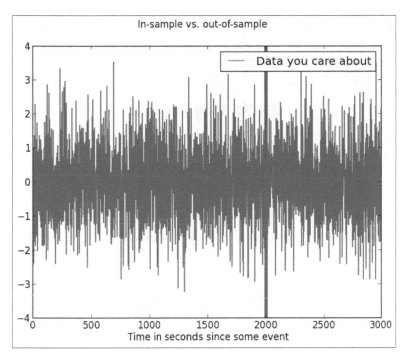

*Figure 6-7. In-sample should come before out-of-sample data in a time series dataset*

One final point on causal modeling and in-sample versus out-of-sample. It is consistent with production code, because we are always acting—in the training and in the out-of-sample simulation—as if we're running our model in production and we're seeing how it performs. Of course we fit our model in sample, so we expect it to perform better there than in production.

Another way to say this is that, once we have a model in production, we will have to make decisions about the future based *only on what we know now* (that's the definition of causal) and we will want to update our model whenever we gather new data. So our coefficients of our model are living organisms that continuously evolve. Just as they should—after all, we're modeling reality, and things really change over time.

# Preparing Financial Data

We often "prepare" the data before putting it into a model. Typically the way we prepare it has to do with the mean or the variance of the data, or sometimes the data after some transformation like the log (and then the mean or the variance of that transformed data). This ends up being a submodel of our model.

---

## Transforming Your Data

Outside of the context of financial data, preparing and transforming data is also a big part of the process. You have a number of possible techniques to choose from to transform your data to better "behave":

- Normalize the data by subtracting the mean and dividing by the standard deviation.

- Alternatively normalize or scale by dividing by the maximum value.

- Take the log of the data.

- Bucket into five evenly spaced buckets; or five evenly distributed buckets (or a number other than five), and create a categorical variable from that.

- Choose a meaningful threshold and transform the data into a new binary variable with value 1, if a data point is greater than or equal to the threshold, and 0 if less than the threshold.

---

Once we have estimates of our mean $\bar{y}$ and variance $\sigma_y^2$, we can normalize the next data point with these estimates just like we do to get from a Gaussian or normal distribution to the standard normal distribution with mean = 0 and standard deviation = 1:

$$y \mapsto \frac{y - \bar{y}}{\sigma_y}$$

Of course we may have other things to keep track of as well to prepare our data, and we might run other submodels of our model. For example, we may choose to consider only the "new" part of something, which is equivalent to trying to predict something like $y_t - y_{t-1}$ instead of $y_t$. Or, we may train a submodel to figure out what part of $y_{t-1}$

---

predicts $y_t$, such as a submodel that is a univariate regression or something.

There are lots of choices here, which will always depend on the situation and the goal you happen to have. Keep in mind, though, that it's all causal, so you have to be careful when you train your overall model how to introduce your next data point and make sure the steps are all in order of time, and that you're never ever cheating and looking ahead in time at data that hasn't happened yet.

In particular, and it happens all the time, one can't normalize by the mean calculuated over the training set. Instead, have a *running estimate of the mean*, which you know at a given moment, and normalize with respect to that.

To see why this is so dangerous, imagine a market crash in the middle of your training set. The mean and variance of your returns are heavily affected by such an event, and doing something as innocuous as a mean estimate translates into anticipating the crash before it happens. Such acausal interference tends to help the model, and could likely make a bad model look good (or, what is more likely, make a model that is pure noise look good).

## Log Returns

In finance, we consider returns on a daily basis. In other words, we care about how much the stock (or future, or index) changes from day to day. This might mean we measure movement from opening on Monday to opening on Tuesday, but the standard approach is to care about closing prices on subsequent trading days.

We typically don't consider percent returns, but rather *log returns*: if $F_t$ denotes a close on day $t$, then the log return that day is defined as $log(F_t / F_{t-1})$, whereas the percent return would be computed as $100((F_t / F_{t-1}) - 1)$. To simplify the discussion, we'll compare log returns to *scaled percent returns*, which is the same as percent returns except without the factor of 100. The reasoning is not changed by this difference in scalar.

There are a few different reasons we use log returns instead of percentage returns. For example, log returns are additive but scaled percent returns aren't. In other words, the five-day log return is the sum of the five one-day log returns. This is often computationally handy.

By the same token, log returns are symmetric with respect to gains and losses, whereas percent returns are biased in favor of gains. So, for example, if our stock goes down by 50%, or has a –0.5 scaled percent gain, and then goes up by 200%, so has a 2.0 scaled percent gain, we are where we started. But working in the same scenarios with log returns, we'd see first a log return of $log(0.5) = -0.301$ followed by a log return of $log(2.0) = 0.301$.

Even so, the two kinds of returns are close to each other for smallish returns, so if we work with short time horizons, like daily or shorter, it doesn't make a huge difference. This can be proven easily: setting $x = F_t / F_{t-1}$, the scaled percent return is $x - 1$ and the log return is $log(x)$, which has the following Taylor expansion:

$$log(x) = \sum_n \frac{(x-1)^n}{n} = (x-1) + (x-1)^2/2 + \cdots$$

In other words, the first term of the Taylor expansion agrees with the percent return. So as long as the second term is small compared to the first, which is usually true for daily returns, we get a pretty good approximation of percent returns using log returns.

Here's a picture of how closely these two functions behave, keeping in mind that when $x = 1$, there's no change in price whatsoever, as shown in Figure 6-8.

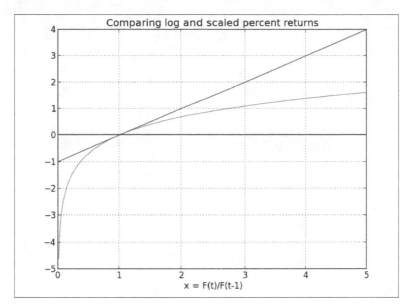

*Figure 6-8. Comparing log and scaled percent returns*

## Example: The S&P Index

Let's work out a toy example. If you start with S&P closing levels as shown in Figure 6-9, then you get the log returns illustrated in Figure 6-10.

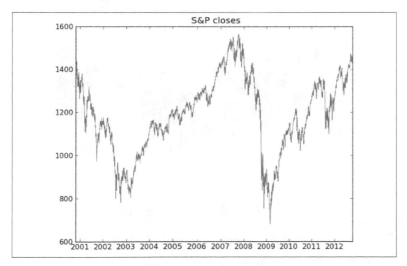

*Figure 6-9. S&P closing levels shown over time*

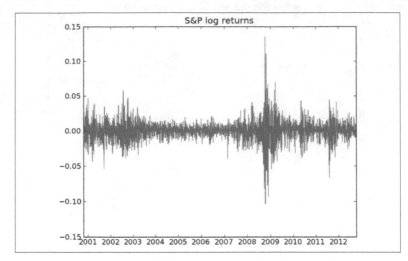

*Figure 6-10. The log of the S&P returns shown over time*

What's that mess? It's crazy volatility caused by the financial crisis. We sometimes (not always) want to account for that volatility by normalizing with respect to it (described earlier). Once we do that we get something like Figure 6-11, which is clearly better behaved.

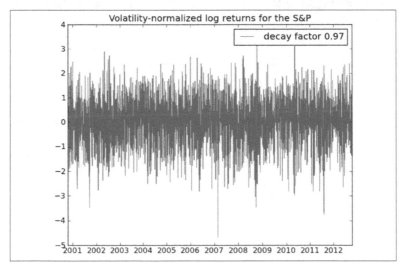

*Figure 6-11. The volatility normalized log of the S&P closing returns shown over time*

# Working out a Volatility Measurement

Once we have our returns defined, we can keep a running estimate of how much we have seen it change recently, which is usually measured as a sample standard deviation, and is called a *volatility estimate*.

A critical decision in measuring the volatility is in choosing a lookback window, which is a length of time in the past we will take our information from. The longer the lookback window is, the more information we have to go by for our estimate. However, the shorter our lookback window, the more quickly our volatility estimate responds to new information. Sometimes you can think about it like this: if a pretty big market event occurs, how long does it take for the market to "forget about it"? That's pretty vague, but it can give one an intuition on the appropriate length of a lookback window. So, for example, it's definitely more than a week, sometimes less than four months. It also depends on how big the event is, of course.

Next we need to decide how we are using the past few days' worth of data. The simplest approach is to take a strictly rolling window, which means we weight each of the previous $n$ days equally and a given day's return is counted for those $n$ days and then drops off the back of a window. The bad news about this easy approach is that a big return will be counted as big until that last moment, and it will completely disappear. This doesn't jive with the ways people forget about things —they usually let information gradually fade from their memories.

For this reason we instead have a continuous lookback window, where we exponentially downweight the older data and we have a concept of the "half-life" of the data. This works out to saying that we scale the impact of the past returns depending on how far back in the past they are, and for each day they get multiplied by some number less than 1 (called the *decay*). For example, if we take the number to be 0.97, then for five days ago we are multiplying the impact of that return by the scalar $0.97^5$. Then we will divide by the sum of the weights, and overall we are taking the weighted average of returns where the weights are just powers of something like 0.97. The "half-life" in this model can be inferred from the number 0.97 using these formulas as $-\ln(2)/\ln(0.97) = 23$.

Now that we have figured out how much we want to weight each previous day's return, we calculate the variance as simply the weighted sum of the squares of the previous returns. Then we take the square root at the end to estimate the volatility.

Note we've just given you a formula that involves all of the previous returns. It's potentially an infinite calculation, albeit with exponentially decaying weights. But there's a cool trick: to actually compute this we only need to keep one running total of the sum so far, and combine it with the new squared return. So we can update our "vol" (as those in the know call volatility) estimate with one thing in memory and one easy weighted average.

First, we are dividing by the sum of the weights, but the weights are powers of some number $s$, so it's a geometric sum and in the limit, the sum is given by $1/(1-s)$.

**Exponential Downweighting**
This technique is called exponential downweighting, a convenient way of compressing the data into a single value that can be updated without having to save the entire dataset.

Next, assume we have the current variance estimate as:

$$V_{old} = (1-s) \cdot \Sigma_i r_i^2 s^i$$

and we have a new return $r_0$ to add to the series. Then it's not hard to show we just want:

$$V_{new} = s \cdot V_{old} + (1-s) \cdot r_0^2$$

Note that we said we would use the sample standard deviation, but the formula for that normally involves removing the mean before taking the sum of squares. Here we ignore the mean, mostly because we are typically taking daily volatility, where the mean (which is hard to anticipate in any case!) is a much smaller factor than the noise, so we can treat it essentially as zero. If we were to measure volatility on a longer time scale such as quarters or years, then we would probably not ignore the mean.

It really matters which downweighting factor you use, as shown in Figure 6-12.

Indeed, we can game a measurement of risk (and people do) by choosing the downweighting factor that minimizes our risk.

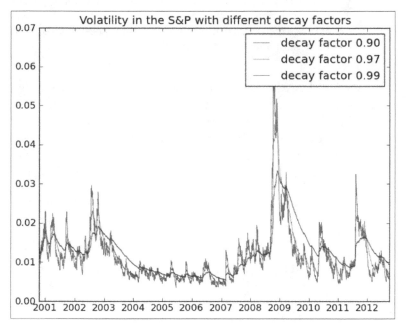

*Figure 6-12. Volatility in the S&P with different decay factors*

## Exponential Downweighting

We've already seen an example of exponential downweighting in the case of keeping a running estimate of the volatility of the returns of the S&P.

The general formula for downweighting some additive running estimate $E$ is simple enough. We weight recent data more than older data, and we assign the downweighting of older data a name $s$ and treat it like a parameter. It is called the *decay*. In its simplest form we get:

$$E_t = s \cdot E_{t-1} + (1-s) \cdot e_t$$

where $e_t$ is the new term.

**Additive Estimates**

We need each of our estimates to be additive (which is why we have a running variance estimate rather than a running standard deviation estimate). If what we're after is a weighted average, say, then we will need to have a running estimate of both numerator and denominator.

If we want to be really careful about smoothness at the beginning (which is more important if we have a few hundred data points or fewer), then we'll actually vary the parameter $s$, via its reciprocal, which we can think of as a kind of half-life. We start with a half-life of 1 and grow it up to the asymptotic "true" half-life $N = 1 / s$. Thus, when we're given a vector $v$ of values $e_t$ indexed by days $t$, we do something like this:

```
true_N = N
this_N_est = 1.0
this_E = 0.0
for e_t in v:
    this_E = this_E * (1-1/this_N_est) + e_t * (1/this_N_est)
    this_N_est = this_N_est*(1-1/true_N) + N * (1/true_N)
```

## The Financial Modeling Feedback Loop

One thing any quantitative person or data scientist needs to understand about financial modeling is that there's a feedback loop. If you find a way to make money, it eventually goes away—sometimes people refer to this as the fact that the market "learns over time."

One way to see this is that, in the end, your model comes down to knowing some price (say) is going to go up in the future, so you buy it before it goes up, you wait, and then you sell it at a profit. But if you think about it, your buying it has actually changed the process, through your market impact, and decreased the signal you were anticipating, at least if the other market players bought it because it looked cheap to them at the previous price; you brought up the price a bit, so you might expect them to buy less in response, which means the overall signal is smaller. Of course, if you only buy one share in anticipation of the increase, your impact is minimal. But if your algorithm works really well, you tend to bet more and more, having a larger and larger impact. Indeed, why would you *not* bet more? You wouldn't. After a while you'd learn the optimal amount you can bet and still make good money, and that optimal amount *is* large enough to have a big impact on the market.

That's how the market learns—it's a combination of a bunch of algorithms anticipating things and making them go away.

The consequence of this learning over time is that the existing signals are very weak. Things that were obvious (in hindsight) with the naked eye in the 1970s are no longer available, because they're all understood and pre-anticipated by the market participants (although new ones might pop into existence).

The bottom line is that, nowadays, we are happy with a 3% correlation for models that have a horizon of 1 day (a "horizon" for your model is how long you expect your prediction to be good). This means not much signal, and lots of noise! Even so, you can still make money if you have such an edge and if your trading costs are sufficiently small.

In particular, lots of the machine learning "metrics of success" for models, such as measurements of precision or accuracy, are not very relevant in this context.

So instead of measuring accuracy, we generally draw a picture to assess models as shown in Figure 6-13, namely of the (cumulative) *PnL* of the model. PnL stands for Profit and Loss and is the day-over-day change (difference, not ratio), or today's value minus yesterday's value.

This generalizes to any model as well—you plot the cumulative sum of the product of *demeaned* forecast and demeaned realized. (A demeaned value is one where the mean's been subtracted.) In other words, you see if your model consistently does better than the "stupidest" model of assuming everything is average.

If you plot this and you drift up and to the right, you're good. If it's too jaggedy, that means your model is taking big bets and isn't stable.

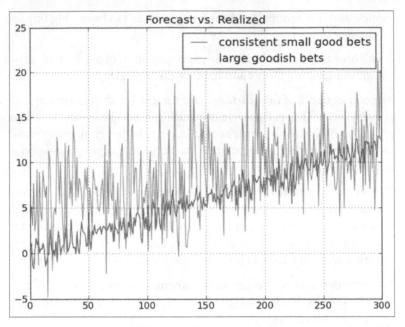

*Figure 6-13. A graph of the cumulative PnLs of two theoretical models*

## Why Regression?

So now we know that in financial modeling, the signal is weak. If you imagine there's some complicated underlying relationship between your information and the thing you're trying to predict, get over knowing what that is—there's too much noise to find it. Instead, think of the function as possibly complicated, but continuous, and imagine you've written it out as a Taylor Series. Then you can't possibly expect to get your hands on anything but the linear terms.

Don't think about using logistic regression, either, because you'd need to be ignoring size, which matters in finance—it matters if a stock went up 2% instead of 0.01%. But logistic regression forces you to have an on/off switch, which would be possible but would lose a lot of information. Considering the fact that we are always in a low-information environment, this is a bad idea.

Note that although we're claiming you probably want to use linear regression in a noisy environment, the actual terms themselves don't have to be linear in the information you have. You can always take products of various terms as x's in your regression. but you're still fitting a linear model in nonlinear terms.

## Adding Priors

One interpretation of priors is that they can be thought of as opinions that are mathematically formulated and incorporated into our models. In fact, we've already encountered a common prior in the form of downweighting old data. The prior can be described as "new data is more important than old data."

Besides that one, we may also decide to consider something like "coefficients vary smoothly." This is relevant when we decide, say, to use a bunch of old values of some time series to help predict the next one, giving us a model like:

$$y = F_t = \alpha_0 + \alpha_1 F_{t-1} + \alpha_2 F_{t-2} + \epsilon$$

which is just the example where we take the last two values of the time series $F$ to predict the next one. We could use more than two values, of course. If we used lots of lagged values, then we could strengthen our prior in order to make up for the fact that we've introduced so many degrees of freedom. In effect, *priors reduce degrees of freedom.*

The way we'd place the prior about the relationship between coefficients (in this case consecutive lagged data points) is by adding a matrix to our covariance matrix when we perform linear regression. See more about this here (*http://goo.gl/gJURp6*).

## A Baby Model

Say we drew a plot in a time series and found that we have strong but fading *autocorrelation* up to the first 40 lags or so as shown in Figure 6-14.

We can calculate autocorrelation when we have time series data. We create a second time series that is the same vector of data shifted by a day (or some fixed time period), and then calculate the correlation between the two vectors.

If we want to predict the next value, we'd want to use the signal that already exists just by knowing the last 40 values. On the other hand, we don't want to do a linear regression with 40 coefficients because that would be way too many degrees of freedom. It's a perfect place for a prior.

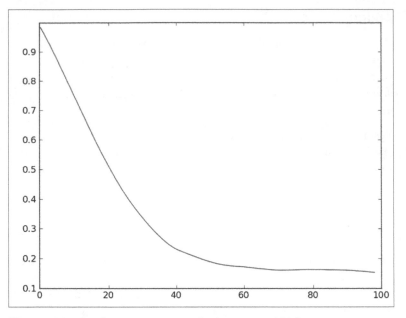

*Figure 6-14. Looking at auto-correlation out to 100 lags*

A good way to think about priors is by adding a term to the function we are seeking to minimize, which measures the extent to which we have a good fit. This is called the "penalty function," and when we have no prior at all, it's simply the sum of the squares of the error:

$$F(\beta) = \Sigma_i\left(y_i - x_i\beta\right)^2 = (y - x\beta)^{\tau}(y - x\beta)$$

If we want to minimize $F$, which we do, then we take its derivative with respect to the vector of coefficients $\beta$, set it equal to zero, and solve for $\beta$—there's a unique solution, namely:

$$\beta = \left(x^{\tau}x\right)^{-1}x^{\tau}y$$

If we now add a standard prior in the form of a penalty term for large coefficients, then we have:

$$F_1(\beta) = \frac{1}{N}\Sigma_i\left(y_i - x_i\beta\right)^2 + \Sigma_j\lambda^2\beta_j^2 = \frac{1}{N}(y - x\beta)^{\tau}(y - x\beta) + (\lambda I\beta)^{\tau}(\lambda I\beta)$$

This can also be solved using calculus, and we solve for $\beta$ to get:

$$\beta_1 = \left(x^\tau x + N \cdot \lambda^2 I\right)^{-1} x^\tau y$$

In other words, adding the penalty term for large coefficients translates into adding a scalar multiple of the identity matrix to the covariance matrix in the closed form solution to $\beta$.

If we now want to add another penalty term that represents a "coefficients vary smoothly" prior, we can think of this as requiring that *adjacent coefficients should be not too different from each other*, which can be expressed in the following penalty function with a new parameter $\mu$ as follows:

$$F_2(\beta) = \frac{1}{N} \sum_i (y_i - x_i\beta)^2 + \sum_j \lambda^2 \beta_j^2 + \sum_j \mu^2 (\beta_j - \beta_{j+1})^2$$

$$= \frac{1}{N} (y - x\beta)^\tau (y - x\beta) + \lambda^2 \beta^\tau \beta + \mu^2 (I\beta - M\beta)^\tau (I\beta - M\beta)$$

where $M$ is the matrix that contains zeros everywhere except on the lower off-diagonals, where it contains 1's. Then $M\beta$ is the vector that results from shifting the coefficients of $\beta$ by one and replacing the last coefficient by 0. The matrix $M$ is called a *shift operator* and the difference $I - M$ can be thought of as a *discrete derivative operator* (see here (*http://goo.gl/2D4LeH*) for more information on discrete calculus).

Because this is the most complicated version, let's look at this in detail. Remembering our vector calculus, the derivative of the scalar function $F_2(\beta)$ with respect to the vector $\beta$ is a vector, and satisfies a bunch of the properties that happen at the scalar level, including the fact that it's both additive and linear and that:

$$\frac{\partial u^\tau \cdot u}{\partial \beta} = 2 \frac{\partial u^\tau}{\partial \beta} u$$

Putting the preceding rules to use, we have:

$$\frac{\partial F_2(\beta)}{\partial \beta} = \frac{1}{N} \frac{\partial (y - x\beta)^\tau (y - x\beta)}{\partial \beta} + \lambda^2 \cdot \frac{\partial \beta^\tau \beta}{\partial \beta} + \mu^2 \cdot \frac{\partial ((I - M)\beta)^\tau ((I - M)\beta)}{\partial \beta}$$

$$= \frac{-2}{N} x^\tau (y - x\beta) + 2\lambda^2 \cdot \beta + 2\mu^2 (I - M)^\tau (I - M)\beta$$

Setting this to 0 and solving for $\beta$ gives us:

$$\beta_2 = \left( x^\tau x + N \cdot \lambda^2 I + N \cdot \mu^2 \cdot (I - M)^\tau (I - M) \right)^{-1} x^\tau y$$

In other words, we have yet another matrix added to our covariance matrix, which expresses our prior that coefficients vary smoothly. Note that the symmetric matrix $(I - M)^\tau (I - M)$ has 1's along its sub- and super-diagonal, but also has 2's along its diagonal. In other words, we need to adjust our $\lambda$ as we adjust our $\mu$ because there is an interaction between these terms.

### Priors and Higher Derivatives

If you want to, you can add a prior about the *second derivative* (or other higher derivatives) as well, by squaring the derivative operator $(I - M)$ (or taking higher powers of it).

So what's the model? Well, remember we will choose an exponential downweighting term $\gamma$ for our data, and we will keep a running estimate of both $x^\tau x$ and $x^\tau y$ as was explained previously. The hyperparameters of our model are then $\gamma$, $\lambda$, and $\mu$. We usually have a sense of how large $\gamma$ should be based on the market, and the other two parameters depend on each other and on the data itself. This is where it becomes an art—you want to optimize to some extent on these choices, without going overboard and overfitting.

## Exercise: GetGlue and Timestamped Event Data

GetGlue kindly provided a dataset for us to explore their data, which contains timestamped events of users checking in and rating TV shows and movies.

The raw data (*http://bit.ly/1aL8XS0*) is on a per-user basis, from 2007 to 2012, and only shows ratings and check-ins for TV shows and movies. It's less than 3% of their total data, and even so it's big, namely 11 GB once it's uncompressed.

Here's some R code to look at the first 10 rows:

```
#
# author: Jared Lander
#
require(rjson)
require(plyr)

# the location of the data
```

```
dataPath <- "http://getglue-data.s3.amazonaws.com/
            getglue_sample.tar.gz"

# build a connection that can decompress the file
theCon <- gzcon(url(dataPath))

# read 10 lines of the data
n.rows <- 10
theLines <- readLines(theCon, n=n.rows)

# check its structure
str(theLines)
# notice the first element is different than the rest
theLines[1]

# use fromJSON on each element of the vector, except the first
theRead <- lapply(theLines[-1], fromJSON)

# turn it all into a data.frame
theData <- ldply(theRead, as.data.frame)

# see how we did
View(theData)
```

Start with these steps:

1.  Load in 1,000 rows of data and spend time looking at it with your eyes. For this entire assignment, work with these 1,000 rows until your code is in good shape. Then you can extend to 100,000 or 1 million rows.

2.  Aggregate and count data. Find answers to the following questions:

    *   How many unique actions can a user take? And how many actions of each type are in this dataset?

    *   How many unique users are in this dataset?

    *   What are the 10 most popular movies?

    *   How many events in this dataset occurred in 2011?

3.  Propose five new questions that you think are interesting and worth investigating.

4.  Investigate/answer your questions.

5.  Visualize. Come up with one visualization that you think captures something interesting about this dataset.

## Exercise: Financial Data

Here's an exercise to help you explore the concepts in this chapter:

1. Get the Data: Go to Yahoo! Finance and download daily data from a stock that has at least eight years of data, making sure it goes from earlier to later. If you don't know how to do it, Google it.

2. Create the time series of daily log returns of the stock price.

3. Just for comparison, do the same for volume data (i.e., create the time series of daily log changes in volume).

4. Next, try to set up a linear regression model that uses the past two returns to predict the next return. Run it and see if you can make any money with it. Try it for both stock returns and volumes. Bonus points if you: do a causal model, normalize for volatility (standard deviation), or put in an exponential decay for old data.

5. Draw the cumulative P&L (forecast × realized) graphs and see if they drift up.

# CHAPTER 7

# Extracting Meaning from Data

How do companies extract meaning from the data they have?

In this chapter we hear from two people with very different approaches to that question—namely, William Cukierski from Kaggle and David Huffaker from Google.

## William Cukierski

Will went to Cornell for a BA in physics and to Rutgers to get his PhD in biomedical engineering. He focused on cancer research, studying pathology images. While working on writing his dissertation, he got more and more involved in Kaggle competitions (more about Kaggle in a bit), finishing very near the top in multiple competitions, and now works for Kaggle.

After giving us some background in data science competitions and crowdsourcing, Will will explain how his company works for the participants in the platform as well as for the larger community.

Will will then focus on *feature extraction* and *feature selection*. Quickly, feature extraction refers to taking the raw dump of data you have and curating it more carefully, to avoid the "garbage in, garbage out" scenario you get if you just feed raw data into an algorithm without enough forethought. Feature selection is the process of constructing a subset of the data or functions of the data to be the predictors or variables for your models and algorithms.

# Background: Data Science Competitions

There is a history in the machine learning community of data science competitions—where individuals or teams compete over a period of several weeks or months to design a prediction algorithm. What it predicts depends on the particular dataset, but some examples include whether or not a given person will get in a car crash, or like a particular film. A training set is provided, an evaluation metric determined up front, and some set of rules is provided about, for example, how often competitors can submit their predictions, whether or not teams can merge into larger teams, and so on.

Examples of machine learning competitions include the annual Knowledge Discovery and Data Mining (KDD) competition, the one-time million-dollar Netflix prize (a competition that lasted two years), and, as we'll learn a little later, Kaggle itself.

Some remarks about data science competitions are warranted. First, data science competitions are part of the data science ecosystem—one of the cultural forces at play in the current data science landscape, and so aspiring data scientists ought to be aware of them.

Second, creating these competitions puts one in a position to codify data science, or define its scope. By thinking about the challenges that they've issued, it provides a set of examples for us to explore the central question of this book: what is data science? This is not to say that we will unquestionably accept such a definition, but we can at least use it as a starting point: what attributes of the existing competitions capture data science, and what aspects of data science are missing?

Finally, competitors in the the various competitions get ranked, and so one metric of a "top" data scientist could be their standings in these competitions. But notice that many top data scientists, especially women, and including the authors of this book, don't compete. In fact, there are few women at the top, and we think this phenomenon needs to be explicitly thought through when we expect top ranking to act as a proxy for data science talent.

**Data Science Competitions Cut Out All the Messy Stuff**
Competitions might be seen as formulaic, dry, and synthetic compared to what you've encountered in normal life. Competitions cut out the messy stuff before you start building models—asking good questions, collecting and cleaning the data, etc.—as well as what happens once you have your model, including visualization and communication. The team of Kaggle data scientists actually spends a lot of time creating the dataset and evaluation metrics, and figuring out what questions to ask, so the question is: while *they're* doing data science, are the contestants?

# Background: Crowdsourcing

There are two kinds of crowdsourcing models. First, we have the *distributive* crowdsourcing model, like Wikipedia, which is for relatively simplistic but large-scale contributions. On Wikipedia, the online encyclopedia, anyone in the world can contribute to the content, and there is a system of regulation and quality control set up by volunteers. The net effect is a fairly high-quality compendium of all of human knowledge (more or less).

Then, there's the singular, focused, difficult problems that Kaggle, DARPA, InnoCentive, and other companies specialize in. These companies issue a challenge to the public, but generally only a set of people with highly specialized skills compete. There is usually a cash prize, and glory or the respect of your community, associated with winning.

Crowdsourcing projects have historically had a number of issues that can impact their usefulness. A couple aspects impact the likelihood that people will participate. First off, many lack an evaluation metric. How do you decide who wins? In some cases, the evaluation method isn't always objective. There might be a subjective measure, where the judges decide your design is bad or they just have different taste. This leads to a high barrier to entry, because people don't trust the evaluation criterion. Additionally, one doesn't get recognition until after they've won or at least ranked highly. This leads to high sunk costs for the participants, which people know about in advance—this can become yet another barrier to entry.

Organizational factors can also hinder success. A competition that is not run well conflates participants with mechanical turks: in other words, they assume the competitors to be somewhat mindless and give bad questions and poor prizes. This precedent is just bad for everyone in that it demoralizes data scientists and doesn't help businesses answer more essential questions that get the most from their data. Another common problem is when the competitions don't chunk the work into bite-sized pieces. Either the question is too big to tackle or too small to be interesting.

Learning from these mistakes, we expect a good competition to have a feasible, interesting question, with an evaluation metric that is transparent and objective. The problem is given, the dataset is given, and the metric of success is given. Moreover, the prizes are established up front.

Let's get a bit of historical context for crowdsourcing, since it is not a new idea. Here are a few examples:

- In 1714, the British Royal Navy couldn't measure longitude, and put out a prize worth $6 million in today's dollars to get help. John Harrison, an unknown cabinetmaker, figured out how to make a clock to solve the problem.

- In 2002, the TV network Fox issued a prize for the next pop solo artist, which resulted in the television show American Idol, where contestants compete in an elimination-round style singing competition.

- There's also the X-prize company (*http://www.xprize.org*), which offers "incentivized prize competitions…to bring about radical breakthroughs for the benefits of humanity, thereby inspiring the formation of new industries and the revitalization of markets." A total of $10 million was offered for the Ansari X-prize, a space competition, and $100 million was invested by contestants trying to solve it. Note this shows that it's not always such an efficient process overall—but on the other hand, it could very well be efficient for the people offering the prize if it gets solved.

# Terminology: Crowdsourcing and Mechanical Turks

These are a couple of terms that have started creeping into the vernacular over the past few years.

Although crowdsourcing—the concept of using many people to solve a problem independently—is not new, the term was only fairly recently coined in 2006. The basic idea is a that a challenge is issued and contestants compete to find the best solution. *The Wisdom of Crowds* was a book written by James Suriowiecki (Anchor, 2004) with the central thesis that, on average, crowds of people will make better decisions than experts, a related phenomenon. It is only under certain conditions (independence of the individuals rather than group-think where a group of people talking to each other can influence each other into wildly incorrect solutions), where groups of people can arrive at the correct solution. And only certain problems are well-suited to this approach.

Amazon Mechanical Turk is an online crowdsourcing service where humans are given tasks. For example, there might be a set of images that need to be labeled as "happy" or "sad." These labels could then be used as the basis of a training set for a supervised learning problem. An algorithm could then be trained on these human-labeled images to automatically label new images. So the central idea of Mechanical Turk is to have humans do fairly routine tasks to help machines, with the goal of the machines then automating tasks to help the humans! Any researcher with a task they need automated can use Amazon Mechanical Turk as long as they provide compensation for the humans. And any human can sign up and be part of the crowdsourcing service, although there are some quality control issues—if the researcher realizes the human is just labeling every other image as "happy" and not actually looking at the images, then the human won't be used anymore for labeling.

Mechanical Turk is an example of artificial artificial intelligence (yes, double up on the "artificial"), in that the humans are helping the machines helping the humans.

# The Kaggle Model

> Being a data scientist is when you learn more and more about more and more, until you know nothing about everything.
>
> — Will Cukierski

Kaggle is a company whose tagline is, "We're making data science a sport." Kaggle forms relationships with companies and with data scientists. For a fee, Kaggle hosts competitions for businesses that essentially want to crowdsource (or leverage the wider data science community) to solve their data problems. Kaggle provides the infrastructure and attracts the data science talent.

They also have in house a bunch of top-notch data scientists, including Will himself. The companies are their paying customers, and they provide datasets and data problems that they want solved. Kaggle crowdsources these problems with data scientists around the world. Anyone can enter. Let's first describe the Kaggle experience for a data scientist and then discuss the customers.

## A Single Contestant

In Kaggle competitions, you are given a training set, and also a test set where the $y$s are hidden, but the $x$s are given, so you just use your model to get your predicted $x$s for the test set and upload them into the Kaggle system to see your evaluation score. This way you don't share your actual code with Kaggle unless you win the prize (and Kaggle doesn't have to worry about which version of Python you're running). Note that even giving out just the $x$s is real information—in particular it tells you, for example, what sizes of $x$s your algorithm should optimize for. Also for the purposes of the competition, there is a third hold-out set that contestants never have access to. You don't see the $x$s or the $y$s—that is used to determine the competition winner when the competition closes.

On Kaggle, the participants are encouraged to submit their models up to five times a day during the competitions, which last a few weeks. As contestants submit their predictions, the Kaggle leaderboard updates immediately to display the contestant's current evaluation metric on the hold-out test set. With a sufficient number of competitors doing this, we see a "leapfrogging" between them as shown in Figure 7-1, where one ekes out a 5% advantage, giving others incentive to work harder. It also establishes a band of accuracy around a problem that you generally don't have—in other words, given no other information, with nobody else working on the problem you're working on, you don't know if your 75% accurate model is the best possible.

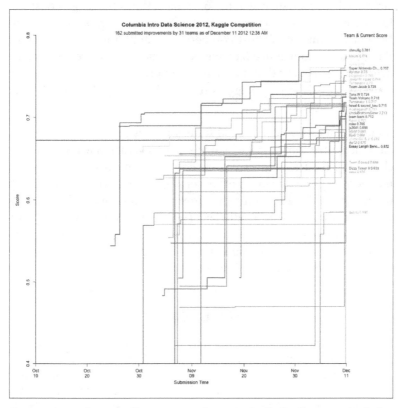

*Figure 7-1. Chris Mulligan, a student in Rachel's class, created this leapfrogging visualization to capture the competition in real time as it progressed throughout the semester*

This leapfrogging effect is good and bad. It encourages people to squeeze out better performing models, possibly at the risk of overfitting, but it also tends to make models much more complicated as they get better. One reason you don't want competitions lasting too long is that, after a while, the only way to inch up performance is to make things ridiculously complicated. For example, the original Netflix Prize lasted two years and the final winning model was too complicated for them to actually put into production.

## Their Customers

So why would companies pay to work with Kaggle? The hole that Kaggle is filling is the following: there's a mismatch between those who need analysis and those with skills. Even though companies desperately need analysis, they tend to hoard data; this is the biggest obstacle for success for those companies that even host a Kaggle competition. Many companies don't host competitions at all, for similar reasons. Kaggle's innovation is that it convinces businesses to share proprietary data with the benefit that their large data problems will be solved for them by crowdsourcing Kaggle's tens of thousands of data scientists around the world.

Kaggle's contests have produced some good results so far. Allstate, the auto insurance company—which has a good actuarial team already—challenged their data science competitors to improve their actuarial model that, given attributes of drivers, approximates the probability of a car crash. The 202 competitors improved Allstate's model by 271% under normalized Gini coefficient (see *http://www.kaggle.com/solutions/casestudies/allstate* for more). Another example includes a company where the prize for competitors was $1,000, and it benefited the company on the order of $100,000.

---

## Is This Fair?

Is it fair to the data scientists already working at the companies that engage with Kaggle? Some of them might lose their job, for example, if the result of the competition is better than the internal model. Is it fair to get people to basically work for free and ultimately benefit a for-profit company? Does it result in data scientists losing their fair market price? Kaggle charges a fee for hosting competitions, and it offers well-defined prizes, so a given data scientist can always choose to not compete. Is that enough?

This seems like it could be a great opportunity for companies, but only while the data scientists of the world haven't realized their value and have extra time on their hands. As soon as they price their skills better they might think twice about working for (almost) free, unless it's for a cause they actually believe in.

---

When Facebook was recently hiring data scientists, they hosted a Kaggle competition, where the prize was an *interview*. There were 422 competitors. We think it's convenient for Facebook to have interviewees for data science positions in such a posture of gratitude for the mere interview. Cathy thinks this distracts data scientists from asking hard questions about what the data policies are and the underlying ethics of the company.

## Kaggle's Essay Scoring Competition

Part of the final exam for the Columbia class was an essay grading contest. The students had to build it, train it, and test it, just like any other Kaggle competition, and group work was encouraged. The details of the essay contest are discussed below, and you access the data at *https://inclass.kaggle.com*.

You are provided access to hand-scored essays so that you can build, train, and test an automatic essay scoring engine. Your success depends upon how closely you can deliver scores to those of human expert graders.

For this competition, there are five essay sets. Each of the sets of essays was generated from a single prompt. Selected essays range from an average length of 150 to 550 words per response. Some of the essays are dependent upon source information and others are not. All responses were written by students ranging in grade levels 7 to 10. All essays were hand graded and were double-scored. Each of the datasets has its own unique characteristics. The variability is intended to test the limits of your scoring engine's capabilities. The data has these columns:

*id*
> A unique identifier for each individual student essay set

*1-5*
> An id for each set of essays

*essay*
> The ASCII text of a student's response

*rater1*
> Rater 1's grade

*rater2*
> Rater 2's grade

*grade*
> Resolved score between the raters

# Thought Experiment: What Are the Ethical Implications of a Robo-Grader?

Will asked students to consider whether they would want their essays automatically graded by an underlying computer algorithm, and what the ethical implications of automated grading would be. Here are some of their thoughts.

*Human graders aren't always fair.*

In the case of doctors, there have been studies where a given doctor is shown the same slide two months apart and gives different diagnoses. We aren't consistent ourselves, even if we think we are. Let's keep that in mind when we talk about the "fairness" of using machine learning algorithms in tricky situations. Machine learning has been used to research cancer, where the stakes are much higher, although there's probably less effort in gaming them.

*Are machines making things more structured, and is this inhibiting creativity?*

Some might argue that people *want* things to be standardized. (It also depends on how much you really care about your grade.) It gives us a consistency that we like. People don't want artistic cars, for example; they want safe cars. Even so, is it wise to move from the human to the machine version of same thing for any given thing? Is there a universal answer or is it a case-by-case kind of question?

*Is the goal of a test to write a good essay or to do well in a standardized test?*

If the latter, you may as well consider a test like a screening: you follow the instructions, and you get a grade depending on how well you follow instructions. Also, the real profit center for standardized testing is, arguably, to sell books to tell you how to take the tests. How does that translate here? One possible way it could translate would be to have algorithms that game the grader algorithms, by building essays that are graded well but are not written by hand. Then we could see education as turning into a war of machines, between the algorithms the students have and the algorithms the teachers have. We'd probably bet on the students in this war.

# Domain Expertise Versus Machine Learning Algorithms

This is a false dichotomy. It isn't either/or. You need both to solve data science problems. However, Kaggle's president Jeremy Howard pissed some domain experts off in a December 2012 *New Scientist* magazine interview with Peter Aldhous, "Specialist Knowledge is Useless and Unhelpful." Here's an excerpt:

PA: What separates the winners from the also-rans?

JH: The difference between the good participants and the bad is the information they feed to the algorithms. You have to decide what to abstract from the data. Winners of Kaggle competitions tend to be curious and creative people. They come up with a dozen totally new ways to think about the problem. The nice thing about algorithms like the random forest is that you can chuck as many crazy ideas at them as you like, and the algorithms figure out which ones work.

PA: That sounds very different from the traditional approach to building predictive models. How have experts reacted?

JH: The messages are uncomfortable for a lot of people. It's controversial because we're telling them: "Your decades of specialist knowledge are not only useless, they're actually unhelpful; your sophisticated techniques are worse than generic methods." It's difficult for people who are used to that old type of science. They spend so much time discussing whether an idea makes sense. They check the visualizations and noodle over it. That is all actively unhelpful.

PA: Is there any role for expert knowledge?

JH: Some kinds of experts are required early on, for when you're trying to work out what problem you're trying to solve. The expertise you need is strategy expertise in answering these questions.

PA: Can you see any downsides to the data-driven, black-box approach that dominates on Kaggle?

JH: Some people take the view that you don't end up with a richer understanding of the problem. But that's just not true: The algorithms tell you what's important and what's not. You might ask why those things are important, but I think that's less interesting. You end up with a predictive model that works. There's not too much to argue about there.

# Feature Selection

The idea of feature selection is identifying the subset of data or transformed data that you want to put into your model.

Prior to working at Kaggle, Will placed highly in competitions (which is how he got the job), so he knows firsthand what it takes to build effective predictive models. Feature selection is not only useful for winning competitions—it's an important part of building statistical models and algorithms in general. Just because you have data doesn't mean it *all* has to go into the model.

For example, it's possible you have many redundancies or correlated variables in your raw data, and so you don't want to include all those variables in your model. Similarly you might want to construct new variables by transforming the variables with a logarithm, say, or turning a continuous variable into a binary variable, before feeding them into the model.

### Terminology: Features, Explanatory Variables, Predictors

Different branches of academia use different terms to describe the same thing. Statisticians say "explanatory variables" or "dependent variables" or "predictors" when they're describing the subset of data that is the input to a model. Computer scientists say "features."

> Feature extraction and selection are the most important but underrated steps of machine learning. Better features are better than better algorithms.
>
> — Will Cukierski

> We don't have better algorithms, we just have more data.
>
> —Peter Norvig
> *Director of Research for Google*

Is it possible, Will muses, that Norvig really wanted to say we have *better features*? You see, more data is sometimes just more data (example: I can record dice rolls until the cows come home, but after a while I'm not getting any value add because my features will converge), but for the more interesting problems that Google faces, the feature landscape is complex/rich/nonlinear enough to benefit from collecting the data that supports those features.

Why? We are getting bigger and bigger datasets, but that's not always helpful. If the number of features is larger than the number of observations, or if we have a sparsity problem, then large isn't necessarily good. And if the huge data just makes it hard to manipulate because of computational reasons (e.g., it can't all fit on one computer, so the data needs to be sharded across multiple machines) without improving our signal, then that's a net negative.

To improve the performance of your predictive models, you want to improve your feature selection process.

## Example: User Retention

Let's give an example for you to keep in mind before we dig into some possible methods. Suppose you have an app that you designed, let's call it Chasing Dragons (shown in Figure 7-2), and users pay a monthly subscription fee to use it. The more users you have, the more money you make. Suppose you realize that only 10% of new users ever come back after the first month. So you have two options to increase your revenue: find a way to increase the retention rate of existing users, or acquire new users. Generally it costs less to keep an existing customer around than to market and advertise to new users. But setting aside that particular cost-benefit analysis of acquistion or retention, let's choose to focus on your user retention situation by building a model that *predicts* whether or not a new user will come back next month based on their behavior this month. You could build such a model in order to *understand* your retention situation, but let's focus instead on building an algorithm that is highly accurate at *predicting*. You might want to use this model to give a free month to users who you predict need the extra incentive to stick around, for example.

A good, crude, simple model you could start out with would be logistic regression, which you first saw back in Chapter 4. This would give you the probability the user returns their second month conditional on their activities in the first month. (There is a rich set of statistical literature called Survival Analysis that could also work well, but that's not necessary in this case—the modeling part isn't what we want to focus on here, it's the data.) You record each user's behavior for the first 30 days after sign-up. You could log every action the user took with timestamps: user clicked the button that said "level 6" at 5:22 a.m., user slew a dragon at 5:23 a.m., user got 22 points at 5:24 a.m., user was shown an ad for deodorant at 5:25 a.m. This would be the data collection phase. Any action the user could take gets recorded.

*Figure 7-2. Chasing Dragons, the app designed by you*

Notice that some users might have thousands of such actions, and other users might have only a few. These would all be stored in time-stamped event logs. You'd then need to process these logs down to a dataset with rows and columns, where each row was a user and each column was a feature. At this point, you shouldn't be selective; you're in the feature generation phase. So your data science team (game designers, software engineers, statisticians, and marketing folks) might sit down and brainstorm features. Here are some examples:

- Number of days the user visited in the first month
- Amount of time until second visit
- Number of points on day $j$ for $j = 1, \ldots, 30$ (this would be 30 separate features)
- Total number of points in first month (sum of the other features)
- Did user fill out Chasing Dragons profile (binary 1 or 0)
- Age and gender of user
- Screen size of device

Use your imagination and come up with as many features as possible. Notice there are redundancies and correlations between these features; that's OK.

# Feature Generation or Feature Extraction

This process we just went through of brainstorming a list of features for Chasing Dragons is the process of *feature generation* or *feature extraction*. This process is as much of an art as a science. It's good to have a domain expert around for this process, but it's also good to use your imagination.

In today's technology environment, we're in a position where we can generate tons of features through logging. Contrast this with other contexts like surveys, for example—you're lucky if you can get a survey respondent to answer 20 questions, let alone hundreds.

But how many of these features are just noise? In this environment, when you can capture a lot of data, not all of it might be actually useful information.

Keep in mind that ultimately you're limited in the features you have access to in two ways: whether or not it's possible to even capture the information, and whether or not it even occurs to you at all to try to capture it. You can think of information as falling into the following buckets:

*Relevant and useful, but it's impossible to capture it.*
> You should keep in mind that there's a lot of information that you're *not* capturing about users—how much free time do they actually have? What other apps have they downloaded? Are they unemployed? Do they suffer from insomnia? Do they have an addictive personality? Do they have nightmares about dragons? Some of this information might be more predictive of whether or not they return next month. There's not much you can do about this, except that it's possible that some of the data you *are* able to capture serves as a proxy by being highly correlated with these unobserved pieces of information: e.g., if they play the game every night at 3 a.m., they might suffer from insomnia, or they might work the night shift.

*Relevant and useful, possible to log it, and you did.*
> Thankfully it occurred to you to log it during your brainstorming session. It's great that you chose to log it, but just because you chose to log it doesn't mean you know that it's relevant or useful, so that's what you'd like your feature selection process to discover.

*Relevant and useful, possible to log it, but you didn't.*
It could be that you didn't think to record whether users uploaded a photo of themselves to their profile, and this action is highly predictive of their likelihood to return. You're human, so sometimes you'll end up leaving out really important stuff, but this shows that your own imagination is a constraint in feature selection. One of the key ways to avoid missing useful features is by doing usability studies (which will be discussed by David Huffaker later on this chapter), to help you think through the user experience and what aspects of it you'd like to capture.

*Not relevant or useful, but you don't know that and log it.*
This is what feature selection is all about—you've logged it, but you don't actually need it and you'd like to be able to know that.

*Not relevant or useful, and you either can't capture it or it didn't occur to you.*
That's OK! It's not taking up space, and you don't need it.

So let's get back to the logistic regression for your game retention prediction. Let $c_i = 1$ if user $i$ returns to use Chasing Dragons any time in the subsequent month. Again this is crude—you could choose the subsequent week or subsequent two months. It doesn't matter. You just first want to get a working model, and then you can refine it.

Ultimately you want your logistic regression to be of the form:

$$\text{logit}\big(P\big(c_i = 1 \big| x_i\big)\big) = \alpha + \beta^\tau \cdot x_i$$

So what should you do? Throw all the hundreds of features you created into one big logistic regression? You could. It's not a terrible thing to do, but if you want to scale up or put this model into production, and get the highest predictive power you can from the data, then let's talk about how you might refine this list of features.

Will found this famous paper by Isabelle Guyon published in 2003 (*http://goo.gl/3dz8Ar*) entitled "An Introduction to Variable and Feature Selection" to be a useful resource. The paper focuses mainly on constructing and selecting subsets of features that are *useful* to build a good predictor. This contrasts with the problem of finding or ranking all potentially relevant variables. In it she studies three categories of feature selection methods: filters, wrappers, and embedded methods.

Keep the Chasing Dragons prediction example in mind as you read on.

# Filters

Filters order possible features with respect to a ranking based on a metric or statistic, such as correlation with the outcome variable. This is sometimes good on a first pass over the space of features, because they then take account of the predictive power of individual features. However, the problem with filters is that you get correlated features. In other words, the filter doesn't care about redundancy. And by treating the features as independent, you're not taking into account possible interactions.

This isn't always bad and it isn't always good, as Isabelle Guyon explains. On the one hand, two redundant features can be more powerful when they are both used; and on the other hand, something that appears useless alone could actually help when combined with another possibly useless-looking feature that an interaction would capture.

Here's an example of a filter: for each feature, run a linear regression with only that feature as a predictor. Each time, note either the p-value or R-squared, and rank order according to the lowest p-value or highest R-squared (more on these two in "Selection criterion" on page 182).

# Wrappers

Wrapper feature selection tries to find subsets of features, of some fixed size, that will do the trick. However, as anyone who has studied combinations and permutations knows, the number of possible size $k$ subsets of $n$ things, called $\binom{n}{k}$, grows exponentially. So there's a nasty opportunity for overfitting by doing this.

There are two aspects to wrappers that you need to consider: 1) selecting an algorithm to use to select features and 2) deciding on a selection criterion or filter to decide that your set of features is "good."

### Selecting an algorithm

Let's first talk about a set of algorithms that fall under the category of *stepwise regression*, a method for feature selection that involves selecting features according to some selection criterion by either adding or subtracting features to a regression model in a systematic way. There are three primary methods of stepwise regression: forward selection, backward elimination, and a combined approach (forward and backward).

*Forward selection*

In forward selection you start with a regression model with no features, and gradually add one feature at a time according to which feature improves the model the most based on a selection criterion. This looks like this: build all possible regression models with a single predictor. Pick the best. Now try all possible models that include that best predictor and a second predictor. Pick the best of those. You keep adding one feature at a time, and you stop when your selection criterion no longer improves, but instead gets worse.

*Backward elimination*

In backward elimination you start with a regression model that includes *all* the features, and you gradually remove one feature at a time according to the feature whose removal makes the biggest improvement in the selection criterion. You stop removing features when removing the feature makes the selection criterion get worse.

*Combined approach*

Most subset methods are capturing some flavor of minimum-redundancy-maximum-relevance. So, for example, you could have a greedy algorithm that starts with the best feature, takes a few more highly ranked, removes the worst, and so on. This a hybrid approach with a filter method.

## Selection criterion

There are a number of selection criteria you could choose from. As a data scientist you have to select which selection criterion to use. Yes! You need a selection criterion to select the selection criterion.

Part of what we wish to impart to you is that in practice, despite the theoretical properties of these various criteria, the choice you make is somewhat arbitrary. One way to deal with this is to try different selection criteria and see how robust your choice of model is. Different selection criterion might produce wildly different models, and it's part of your job to decide what to optimize for and why:

*R-squared*

Given by the formula $R^2 = 1 - \frac{\Sigma_i (y_i - \hat{y_i})^2}{\Sigma_i (y_i - \bar{y})^2}$, it can be interpreted as the proportion of variance explained by your model.

*p-values*

In the context of regression where you're trying to estimate coefficients (the $\beta$s), to think in terms of p-values, you make an assumption of there being a *null hypothesis* that the $\beta$s are zero. For any given $\beta$, the p-value captures the probability of observing the data that you observed, and obtaining the test-statistic (in this case the estimated $\hat{\beta}$) that you got *under the null hypothesis*. Specifically, if you have a low p-value, it is highly unlikely that you would observe such a test-statistic if the null hypothesis actually held. This translates to meaning that (with some confidence) the coefficient is highly likely to be non-zero.

*AIC (Akaike Information Criterion)*

Given by the formula $2k - 2ln(L)$, where $k$ is the number of parameters in the model and $ln(L)$ is the "maximized value of the log likelihood." The goal is to minimize AIC.

*BIC (Bayesian Information Criterion)*

Given by the formula $k * ln(n) - 2ln(L)$, where $k$ is the number of parameters in the model, $n$ is the number of observations (data points, or users), and $ln(L)$ is the maximized value of the log likelihood. The goal is to minimize BIC.

*Entropy*

This will be discussed more in "Embedded Methods: Decision Trees" on page 184.

## In practice

As mentioned, stepwise regression is exploring a large space of all possible models, and so there is the danger of overfitting—it will often fit much better in-sample than it does on new out-of-sample data.

You don't have to retrain models at each step of these approaches, because there are fancy ways to see how your objective function (aka selection criterion) changes as you change the subset of features you are trying out. These are called "finite differences" and rely essentially on Taylor Series expansions of the objective function.

One last word: if you have a domain expert on hand, don't go into the machine learning rabbit hole of feature selection unless you've tapped into your expert completely!

## Embedded Methods: Decision Trees

Decision trees have an intuitive appeal because outside the context of data science in our every day lives, we can think of breaking big decisions down into a series of questions. See the decision tree in Figure 7-3 about a college student facing the very important decision of how to spend her time.

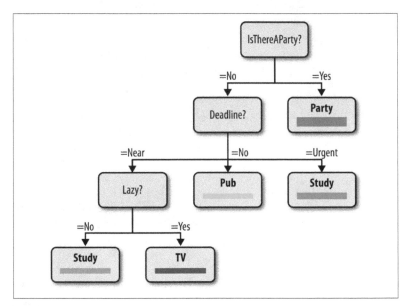

*Figure 7-3. Decision tree for college student, aka the party tree (taken with permission from Stephen Marsland's book, Machine Learning: An Algorithmic Perspective [Chapman and Hall/CRC])*

This decision is actually dependent on a bunch of factors: whether or not there are any parties or deadlines, how lazy the student is feeling, and what they care about most (parties). The interpretability of decision trees is one of the best features about them.

In the context of a data problem, a decision tree is a classification algorithm. For the Chasing Dragons example, you want to classify users as "Yes, going to come back next month" or "No, not going to come back next month." This isn't really a decision in the colloquial sense, so don't let that throw you. You know that the class of any given user is dependent on many factors (number of dragons the user slew, their age, how many hours they already played the game). And you want to break it down based on the data you've collected. But how do you construct decision trees from data and what mathematical properties can you expect them to have?

Ultimately you want a tree that is something like Figure 7-4.

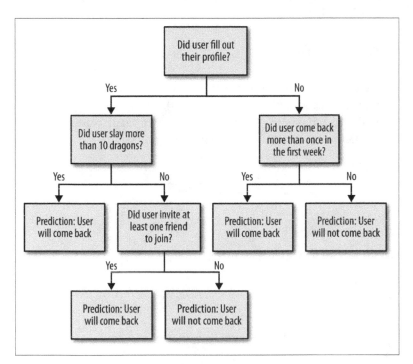

*Figure 7-4. Decision tree for Chasing Dragons*

But you want this tree to be based on data and not just what you feel like. Choosing a feature to pick at each step is like playing the game 20 Questions really well. You take whatever *the most informative thing* is first. Let's formalize that—we need a notion of "informative."

For the sake of this discussion, assume we break compound questions into multiple yes-or-no questions, and we denote the answers by "0" or "1." Given a random variable $X$, we denote by $p(X = 1)$ and $p(X = 0)$ the probability that $X$ is true or false, respectively.

## Entropy

To quantify what is the most "informative" feature, we define entropy—effectively a measure for how mixed up something is—for $X$ as follows:

$$H(X) = -p(X=1)\ log_2(p(X=1)) - p(X=0)\ log_2(p(X=0))$$

Note when $p(X=1)=0$ or $p(X=0)=0$, the entropy vanishes, consistent with the fact that:

$$\lim_{t \to 0} t \cdot log(t) = 0$$

In particular, if either option has probability zero, the entropy is 0. Moreover, because $p(X=1) = 1 - p(X=0)$, the entropy is symmetric about 0.5 and maximized at 0.5, which we can easily confirm using a bit of calculus. Figure 7-5 shows a picture of that.

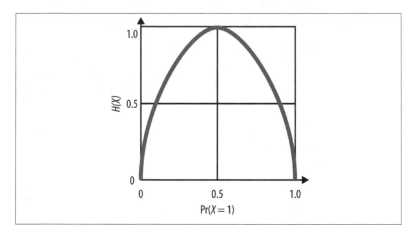

*Figure 7-5. Entropy*

Mathematically, we kind of get this. But what does it mean in words, and why are we calling it entropy? Earlier, we said that entropy is a measurement of how mixed up something is.

So, for example, if $X$ denotes the event of a baby being born a boy, we'd expect it to be true or false with probability close to 1/2, which corresponds to high entropy, i.e., the bag of babies from which we are selecting a baby is highly mixed.

But if $X$ denotes the event of a rainfall in a desert, then it's low entropy. In other words, the bag of day-long weather events is not highly mixed in deserts.

Using this concept of entropy, we will be thinking of $X$ as the *target* of our model. So, $X$ could be the event that someone buys something on our site. We'd like to know which attribute of the user will tell us the most information about this event $X$. We will define the *information gain*, denoted $IG(X,a)$, for a given attribute $a$, as the entropy we *lose* if we know the value of that attribute:

$$IG(X,a) = H(X) - H(X|a)$$

To compute this we need to define $H(X|a)$. We can do this in two steps. For any actual value $a_0$ of the attribute $a$ we can compute the *specific conditional entropy* $H(X|a=a_0)$ as you might expect:

$$H(X|a=a_0) = -p(X=1|a=a_0)\ log_2(p(X=1|a=a_0)) - $$
$$p(X=0|a=a_0)\ log_2(p(X=0|a=a_0))$$

and then we can put it all together, for all possible values of $a$, to get the *conditional entropy* $H(X|a)$:

$$H(X|a) = \Sigma_{a_i}\ p(a=a_i) \cdot H(X|a=a_i)$$

In words, the conditional entropy asks: how mixed is our bag really if we *know* the value of attribute $a$? And then information gain can be described as: how much information do we learn about $X$ (or how much entropy do we lose) once we know $a$?

Going back to how we use the concept of entropy to build decision trees: it helps us decide what feature to split our tree on, or in other words, what's the most informative question to ask?

## The Decision Tree Algorithm

You build your decision tree iteratively, starting at the root. You need an algorithm to decide which attribute to split on; e.g., which node should be the next one to identify. You choose that attribute in order to *maximize information gain*, because you're getting the most bang for your buck that way. You keep going until all the points at the end are in the same class or you end up with no features left. In this case, you take the majority vote.

Often people "prune the tree" afterwards to avoid overfitting. This just means cutting it off below a certain depth. After all, by design, the algorithm gets weaker and weaker as you build the tree, and it's well known that if you build the entire tree, it's often less accurate (with new data) than if you prune it.

This is an example of an embedded feature selection algorithm. (Why embedded?) You don't need to use a filter here because the information gain method is doing your feature selection for you.

Suppose you have your Chasing Dragons dataset. Your outcome variable is *Return*: a binary variable that captures whether or not the user returns next month, and you have tons of predictors. You can use the R library rpart and the function rpart, and the code would look like this:

```
# Classification Tree with rpart
library(rpart)

# grow tree
model1 <- rpart(Return ~ profile + num_dragons +
num_friends_invited + gender + age +
num_days, method="class", data=chasingdragons)

printcp(model1) # display the results
plotcp(model1) # visualize cross-validation results
summary(model1) # detailed summary of thresholds picked to
transform to binary

# plot tree
plot(model1, uniform=TRUE,
        main="Classification Tree for Chasing Dragons")
text(model1, use.n=TRUE, all=TRUE, cex=.8)
```

## Handling Continuous Variables in Decision Trees

Packages that already implement decision trees can handle continuous variables for you. So you can provide continuous features, and it will determine an optimal threshold for turning the continuous variable into a binary predictor. But if *you* are building a decision tree algorithm yourself, then in the case of continuous variables, you need to determine the correct threshold of a value so that it can be thought of as a binary variable. So you could partition a user's number of dragon slays into "less than 10" and "at least 10," and you'd be getting back to the binary variable case. In this case, it takes some extra work to decide on the information gain because it depends on the threshold as well as the feature.

In fact, you could think of the decision of where the threshold should live as a separate submodel. It's possible to optimize to this choice by maximizing the entropy on individual attributes, but that's not clearly the best way to deal with continuous variables. Indeed, this kind of question can be as complicated as feature selection itself—instead of a single threshold, you might want to create bins of the value of your attribute, for example. What to do? It will always depend on the situation.

# Surviving the Titanic

For fun, Will pointed us to this decision tree for surviving on the Titanic (*http://goo.gl/YsyWJW*) on the BigML website. The original data is from the Encyclopedia Titanica (*http://www.encyclopedia-titanica.org/*)–source code and data are available there. Figure 7-6 provides just a snapshot of it, but if you go to the site, it is interactive.

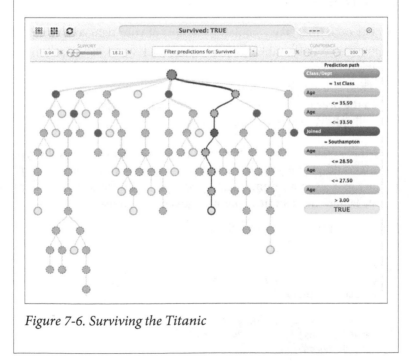

*Figure 7-6. Surviving the Titanic*

# Random Forests

Let's turn to another algorithm for feature selection. Random forests generalize decision trees with *bagging*, otherwise known as *bootstrap aggregating*. We will explain bagging in more detail later, but the effect of using it is to make your models more accurate and more robust, but at the cost of interpretability—random forests are notoriously difficult to understand. They're conversely easy to specify, with two hyperparameters: you just need to specify the number of trees you want in your forest, say $N$, as well as the number of features to randomly select for each tree, say $F$.

Before we get into the weeds of the random forest algorithm, let's review bootstrapping. A *bootstrap sample* is a sample with replacement, which means we might sample the same data point more than once. We usually take to the sample size to be 80% of the size of the entire (training) dataset, but of course this parameter can be adjusted depending on circumstances. This is technically a third hyperparameter of our random forest algorithm.

Now to the algorithm. To construct a random forest, you construct $N$ decision trees as follows:

1. For each tree, take a bootstrap sample of your data, and for each node you randomly select $F$ features, say 5 out of the 100 total features.

2. Then you use your entropy-information-gain engine as described in the previous section to decide which among those features you will split your tree on at each stage.

Note that you could decide beforehand how deep the tree should get, or you could prune your trees after the fact, but you typically *don't* prune the trees in random forests, because a great feature of random forests is that they can incorporate idiosyncratic noise.

The code for this would look like:

```
# Author: Jared Lander
#
# we will be using the diamonds data from ggplot
require(ggplot2)

# load and view the diamonds data
data(diamonds)
head(diamonds)
```

```
# plot a histogram with a line marking $12,000
ggplot(diamonds) + geom_histogram(aes(x=price)) +
geom_vline(xintercept=12000)

# build a TRUE/FALSE variable indicating if the price is above
# our threshold
diamonds$Expensive <- ifelse(diamonds$price >= 12000, 1, 0)
head(diamonds)

# get rid of the price column
diamonds$price <- NULL

## glmnet
require(glmnet)
# build the predictor matrix, we are leaving out the last
# column which is our response
x <- model.matrix(~., diamonds[, -ncol(diamonds)])
# build the response vector
y <- as.matrix(diamonds$Expensive)
# run the glmnet
system.time(modGlmnet <- glmnet(x=x, y=y, family="binomial"))
# plot the coefficient path
plot(modGlmnet, label=TRUE)

# this illustrates that setting a seed allows you to recreate
# random results, run them both a few times
set.seed(48872)
sample(1:10)

## decision tree
require(rpart)
# fire a simple decision tree
modTree <- rpart(Expensive ~ ., data=diamonds)
# plot the splits
plot(modTree)
text(modTree)


## bagging (or bootstrap aggregating)
require(boot)
mean(diamonds$carat)
sd(diamonds$carat)
# function for bootstrapping the mean
boot.mean <- function(x, i)
{
    mean(x[i])
}
# allows us to find the variability of the mean
boot(data=diamonds$carat, statistic=boot.mean, R=120)
require(adabag)
```

```
modBag <- bagging(formula=Species ~ ., iris, mfinal=10)

## boosting
require(mboost)
system.time(modglmBoost <- glmboost(as.factor(Expensive) ~ .,
        data=diamonds, family=Binomial(link="logit")))
summary(modglmBoost)
?blackboost

## random forests
require(randomForest)
system.time(modForest <- randomForest(Species ~ ., data=iris,
        importance=TRUE, proximity=TRUE))
```

## Criticisms of Feature Selection

Let's address a common criticism of feature selection. Namely, it's no
better than data dredging. If we just take whatever answer we get that
correlates with our target, however far afield it is, then we could end
up thinking that Bangladeshi butter production predicts the S&P.
Generally we'd like to first curate the candidate features at least to
some extent. Of course, the more observations we have, the less we
need to be concerned with spurious signals.

There's a well-known bias-variance tradeoff: a model is "high bias" if
it's is too simple (the features aren't encoding enough information).
In this case, lots more data doesn't improve our model. On the other
hand, if our model is too complicated, then "high variance" leads to
overfitting. In this case we want to reduce the number of features we
are using.

## User Retention: Interpretability Versus Predictive Power

So let's say you built the decision tree, and that it predicts quite well.
But should you interpret it? Can you try to find meaning in it?

It could be that it basically tells you "the more the user plays in the first
month, the more likely the user is to come back next month," which
is kind of useless, and this kind of thing happens when you're doing
analysis. It feels circular–*of course* the more they like the app now, the
more likely they are to come back. But it could also be that it tells you
that showing them ads in the first five minutes *decreases* their chances
of coming back, but it's OK to show ads after the first hour, and this

would give you some insight: don't show ads in the first five minutes! Now to study this more, you really would want to do some A/B testing (see Chapter 11), but this initial model and feature selection would help you prioritize the types of tests you might want to run.

It's also worth noting that features that have to do with the user's behavior (user played 10 times this month) are qualitatively different than features that have to do with your behavior (you showed 10 ads, and you changed the dragon to be red instead of green). There's a causation/correlation issue here. If there's a correlation of getting a high number of points in the first month with returning to play next month, does that mean if you just *give* users a high number of points this month without them playing at all, they'll come back? No! It's not the number of points that caused them to come back, it's that they're really into playing the game (a confounding factor), which correlates with both their coming back *and* their getting a high number of points. You therefore would want to do feature selection with *all* variables, but then focus on the ones you can do something about (e.g., show fewer ads) conditional on user attributes.

# David Huffaker: Google's Hybrid Approach to Social Research

David's focus is on the effective marriages of both qualitative and quantitative research, and of big and little data. Large amounts of big quantitative data can be more effectively extracted if you take the time to think on a small scale carefully first, and then leverage what you learned on the small scale to the larger scale. And vice versa, you might find patterns in the large dataset that you want to investigate by digging in deeper by doing intensive usability studies with a handful of people, to add more color to the phenomenon you are seeing, or verify interpretations by connecting your exploratory data analysis on the large dataset with relevant academic literature.

David was one of Rachel's colleagues at Google. They had a successful collaboration—starting with complementary skill sets, an explosion of goodness ensued when they were put together to work on Google+ (Google's social layer) with a bunch of other people, especially software engineers and computer scientists. David brings a social scientist perspective to the analysis of social networks. He's strong in quantitative methods for understanding and analyzing online social behavior. He got a PhD in media, technology, and society from Northwestern

University. David spoke about Google's approach to social research to encourage the class to think in ways that connect the qualitative to the quantitative, and the small-scale to the large-scale.

Google does a good job of putting people together. They blur the lines between research and development. They even wrote about it in this July 2012 position paper: Google's Hybrid Approach to Research (*http://goo.gl/ejtPw2*). Their researchers are embedded on product teams. The work is iterative, and the engineers on the team strive to have near-production code from day 1 of a project. They leverage engineering infrastructure to deploy experiments to their mass user base and to rapidly deploy a prototype at scale. Considering the scale of Google's user base, redesign as they scale up is not a viable option. They instead do experiments with smaller groups of users.

## Moving from Descriptive to Predictive

David suggested that as data scientists, we consider how to move into an experimental design so as to move to a causal claim between variables rather than a descriptive relationship. In other words, our goal is to move from the descriptive to the predictive.

As an example, he talked about the genesis of the "circle of friends" feature of Google+. Google knows people want to selectively share; users might send pictures to their family, whereas they'd probably be more likely to send inside jokes to their friends. Google came up with the idea of circles, but it wasn't clear if people would use them. How can Google answer the question of whether people will use circles to organize their social network? It's important to know what motivates users when they decide to share.

Google took a *mixed-method approach*, which means they used multiple methods to triangulate on findings and insights. Some of their methods were small and qualitative, some of them larger and quantitative.

Given a random sample of 100,000 users, they set out to determine the popular names and categories of names given to circles. They identified 168 active users who filled out surveys and they had longer interviews with 12. The depth of these interviews was weighed against selection bias inherent in finding people that are willing to be interviewed.

They found that the majority were engaging in selective sharing, that most people used circles, and that the circle names were most often work-related or school-related, and that they had elements of a strong-link ("epic bros") or a weak-link ("acquaintances from PTA").

They asked the survey participants why they share content. The answers primarily came in three categories. First, the desire to share about oneself: personal experiences, opinions, etc. Second, discourse: people want to participate in a conversation. Third, evangelism: people like to spread information.

Next they asked participants why they choose their audiences. Again, three categories: first, privacy—many people were public or private by default. Second, relevance: they wanted to share only with those who may be interested, and they don't want to pollute other people's data stream. Third, distribution: some people just want to maximize their potential audience.

The takeaway from this study was that people do enjoy selectively sharing content, depending on context and the audience. So Google has to think about designing features for the product around content, context, and audience. They want to not just keep these findings in mind for design, but also when they do data science at scale.

---

## Thought Experiment: Large-Scale Network Analysis

We'll dig more into network analysis in Chapter 10 with John Kelly. But for now, think about how you might take the findings from the Google+ usability studies and explore selectively sharing content on a massive scale using data. You can use large data and look at connections between actors like a graph. For Google+, the users are the nodes and the edges (directed) are "in the same circle." Think about what data you would want to log, and then how you might test some of the hypotheses generated from speaking with the small group of engaged users.

As data scientists, it can be helpful to think of different structures and representations of data, and once you start thinking in terms of networks, you can see them everywhere.

---

Other examples of networks:

- The nodes are users in Second Life (*http://secondlife.com/*), and the edges correspond to interactions between users. Note there is more than one possible way for players to interact in this game, leading to potentially different kinds of edges.
- The nodes are websites, the (directed) edges are links.
- Nodes are theorems, directed edges are dependencies (*http://www.math.columbia.edu/~dejong/plaatje.png*).

## Social at Google

As you may have noticed, "social" is a layer across all of Google. Search now incorporates this layer: if you search for something you might see that your friend "+1"ed it. This is called a *social annotation*. It turns out that people care more about annotation when it comes from someone with domain expertise rather than someone you're very close to. So you might care more about the opinion of a wine expert at work than the opinion of your mom when it comes to purchasing wine.

Note that sounds obvious but if you started the other way around, asking who you'd trust, you might start with your mom. In other words, "close ties"–even if you can determine those—are not the best feature to rank annotations. But that begs the question, what is? Typically in a situation like this, data scientists might use click-through rate, or how long it takes to click.

In general you need to always keep in mind a quantitative metric of success. This defines success for you, so you have to be careful.

## Privacy

Human-facing technology has thorny issues of privacy, which makes stuff hard. Google conducted a survey of how people felt uneasy about content. They asked, how does it affect your engagement? What is the nature of your privacy concerns?

Turns out there's a strong correlation between privacy concern and low engagement, which isn't surprising. It's also related to how well you understand what information is being shared, and the question of when you post something, where it goes, and how much control you have over it. When you are confronted with a huge pile of complicated settings, you tend to start feeling passive.

The survey results found broad categories of concern as follows:

*Identity theft*
- Financial loss

*Digital world*
- Access to personal data
- Really private stuff I searched on
- Unwanted spam
- Provocative photo (oh *&!$ my boss saw that)
- Unwanted solicitation
- Unwanted ad targeting

*Physical world*
- Offline threats/harassment
- Harm to my family
- Stalkers
- Employment risks

## Thought Experiment: What Is the Best Way to Decrease Concern and Increase Understanding and Control?

So given users' understandable concerns about privacy, students in Rachel's class brainstormed some potential solutions that Google could implement (or that anyone dealing with user-level data could consider).

Possibilities:

- You could write and post a manifesto of your data policy. Google tried that, but it turns out nobody likes to read manifestos.
- You could educate users on your policies a la the Netflix feature "because you liked this, we think you might like this." But it's not always so easy to explain things in complicated models.
- You could simply get rid of all stored data after a year. But you'd still need to explain that you do that.

Maybe we could rephrase the question: how do you design privacy settings to make it easier for people? In particular, how do you make it transparent? Here are some ideas along those lines:

- Make a picture or graph of where data is going.
- Give people a privacy switchboard.
- Provide access to quick settings.
- Make the settings you show people categorized by "things you don't have a choice about" versus "things you do" for the sake of clarity.
- Best of all, you could make reasonable default setting so people don't have to worry about it.

David left us with these words of wisdom: as you move forward and have access to Big Data, you really should complement them with qualitative approaches. Use mixed methods to come to a better understanding of what's going on. Qualitative surveys can really help.

# CHAPTER 8

# Recommendation Engines: Building a User-Facing Data Product at Scale

Recommendation engines, also called recommendation systems, are the quintessential data product and are a good starting point when you're explaining to non–data scientists what you do or what data science really is. This is because many people have interacted with recommendation systems when they've been suggested books on Amazon.com or gotten recommended movies on Netflix. Beyond that, however, they likely have not thought much about the engineering and algorithms underlying those recommendations, nor the fact that their behavior when they buy a book or rate a movie is generating data that then feeds back into the recommendation engine and leads to (hopefully) improved recommendations for themselves and other people.

Aside from being a clear example of a product that literally uses data as its fuel, another reason we call recommendation systems "quintessential" is that building a solid recommendation system end-to-end requires an understanding of linear algebra *and* an ability to code; it also illustrates the challenges that Big Data poses when dealing with a problem that makes intuitive sense, but that can get complicated when implementing its solution at scale.

In this chapter, Matt Gattis walks us through what it took for him to build a recommendation system for Hunch.com—including why he made certain decisions, and how he thought about trade-offs between various algorithms when building a large-scale engineering system and infrastructure that powers a user-facing product.

Matt graduated from MIT in CS, worked at SiteAdvisor, and co-founded Hunch (*http://hunch.com/*) as its CTO. Hunch is a website that gives you recommendations of any kind. When they started out, it worked like this: they'd ask people a bunch of questions (people seem to love answering questions), and then someone could ask the engine questions like, "What cell phone should I buy?" or, "Where should I go on a trip?" and it would give them advice. They use machine learning to give better and better advice. Matt's role there was doing the R&D for the underlying recommendation engine.

At first, they focused on trying to make the questions as fun as possible. Then, of course, they saw things needing to be asked that would be extremely informative as well, so they added those. Then they found that they could ask merely 20 questions and then predict the rest of them with 80% accuracy. There were questions that you might imagine and some that are surprising, like whether people were competitive versus uncompetitive, introverted versus extroverted, thinking versus perceiving, etc.—not unlike MBTI.

Eventually Hunch expanded into more of an API model where they crawl the Web for data rather than asking people direct questions. The service can also be used by third parties to personalize content for a given site—a nice business proposition that led to eBay acquiring Hunch.com.

Matt has been building code since he was a kid, so he considers software engineering to be his strong suit. Hunch requires cross-domain experience so he doesn't consider himself a domain expert in any focused way, except for recommendation systems themselves.

The best quote Matt gave us was this: "Forming a data team is kind of like planning a heist." He means that you need people with all sorts of skills, and that one person probably can't do everything by herself. (Think Ocean's Eleven, but sexier.)

# A Real-World Recommendation Engine

Recommendation engines are used all the time—what movie would you like, knowing other movies you liked? What book would you like, keeping in mind past purchases? What kind of vacation are you likely to embark on, considering past trips?

There are plenty of different ways to go about building such a model, but they have very similar feels if not implementation. We're going to show you how to do one relatively simple but complete version in this chapter.

To set up a recommendation engine, suppose you have *users*, which form a set $U$; and you have *items* to recommend, which form a set $V$. As Kyle Teague told us in Chapter 6, you can denote this as a bipartite graph (shown again in Figure 8-1) if each user and each item has a node to represent it—there is a line from a user to an item if that user has expressed an opinion about that item. Note they might not always *love* that item, so the edges could have weights: they could be positive, negative, or on a continuous scale (or discontinuous, but many-valued like a star system). The implications of this choice can be heavy but we won't delve too deep here—for us they are numeric ratings.

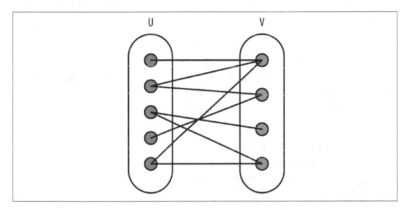

*Figure 8-1. Bipartite graph with users and items (television shows) as nodes*

Next up, you have training data in the form of some preferences—you know some of the opinions of some of the users on some of the items. From those training data, you want to predict other preferences for your users. That's essentially the output for a recommendation engine.

You may also have metadata on users (i.e., they are male or female, etc.) or on items (the color of the product). For example, users come to your website and set up accounts, so you may know each user's gender, age, and preferences for up to three items.

You represent a given user as a vector of features, sometimes including only metadata—sometimes including only preferences (which would lead to a sparse vector because you don't know all the user's opinions)—and sometimes including both, depending on what you're doing with the vector. Also, you can sometimes bundle all the user vectors together to get a big user matrix, which we call $U$, through abuse of notation.

## Nearest Neighbor Algorithm Review

Let's review the nearest neighbor algorithm (discussed in Chapter 3): if you want to predict whether user A likes something, you look at a user B *closest* to user A who has an opinion, then you assume A's opinion is the same as B's. In other words, once you've identified a similar user, you'd then find something that user A hadn't rated (which you'd assume meant he hadn't ever seen that movie or bought that item), but that user B *had* rated and liked and use that as your recommendation for user A.

As discussed in Chapter 3, to implement this you need a metric so you can measure distance. One example when the opinions are binary: Jaccard distance, i.e., 1–(the number of things they both like divided by the number of things either of them likes). Other examples include cosine similarity or Euclidean distance.

**Which Metric Is Best?**
You might get a different answer depending on which metric you choose. But that's a good thing. Try out lots of different distance functions and see how your results change and think about why.

## Some Problems with Nearest Neighbors

So you *could* use nearest neighbors; it makes some intuitive sense that you'd want to recommend items to people by finding similar people and using those people's opinions to generate ideas and recommendations. But there are a number of problems nearest neighbors poses. Let's go through them:

*Curse of dimensionality*
    There are too many dimensions, so the closest neighbors are too far away from each other to realistically be considered "close."

## Overfitting

Overfitting is also a problem. So one guy is closest, but that could be pure noise. How do you adjust for that? One idea is to use k-NN, with, say, k=5 rather than k=1, which increases the noise.

## Correlated features

There are tons of features, moreover, that are highly correlated with each other. For example, you might imagine that as you get older you become more conservative. But then counting both age and politics would mean you're double counting a single feature in some sense. This would lead to bad performance, because you're using redundant information and essentially placing double the weight on some variables. It's preferable to build in an understanding of the correlation and project onto smaller dimensional space.

## Relative importance of features

Some features are more informative than others. Weighting features may therefore be helpful: maybe your age has nothing to do with your preference for item 1. You'd probably use something like covariances to choose your weights.

## Sparseness

If your vector (or matrix, if you put together the vectors) is too sparse, or you have lots of missing data, then most things are unknown, and the Jaccard distance means nothing because there's no overlap.

## Measurement errors

There's measurement error (also called reporting error): people may lie.

## Computational complexity

There's a calculation cost—computational complexity.

## Sensitivity of distance metrics

Euclidean distance also has a scaling problem: distances in age outweigh distances for other features if they're reported as 0 (for don't like) or 1 (for like). Essentially this means that raw euclidean distance doesn't make much sense. Also, old and young people might think one thing but middle-aged people something else. We seem to be assuming a linear relationship, but it may not exist. Should you be binning by age group instead, for example?

*Preferences change over time*

> User preferences may also change over time, which falls outside the model. For example, at eBay, they might be buying a printer, which makes them only want ink for a short time.

*Cost to update*

> It's also expensive to update the model as you add more data.

The biggest issues are the first two on the list, namely overfitting and the curse of dimensionality problem. How should you deal with them? Let's think back to a method you're already familiar with—linear regression—and build up from there.

## Beyond Nearest Neighbor: Machine Learning Classification

We'll first walk through a simplification of the actual machine learning algorithm for this—namely we'll build a separate linear regression model for each item. With each model, we could then predict for a given user, knowing their attributes, whether they would like the item corresponding to that model. So one model might be for predicting whether you like *Mad Men* and another model might be for predicting whether you would like Bob Dylan.

Denote by $f_{i,j}$ user $i$'s stated preference for item $j$ if you have it (or user $i$'s attribute, if item $j$ is a metadata item like age or is_logged_in). This is a subtle point but can get a bit confusing if you don't internalize this: you are treating metadata here *also* as if it's an "item." We mentioned this before, but it's OK if you didn't get it—hopefully it will click more now. When we said we could predict what you might like, we're also saying we could use this to predict your *attribute*; i.e., if we didn't *know* if you were a male/female because that was missing data or we had never asked you, we might be able to predict that.

To let this idea settle even more, assume we have three numeric attributes for each user, so we have $f_{i,1}, f_{i,2}$, and $f_{i,3}$. Then to guess user $i$'s preference on a new item (we temporarily denote this estimate by $p_i$) we can look for the best choice of $\beta_k$ so that:

$$p_i = \beta_1 f_{i,1} + \beta_2 f_{i,2} + \beta_3 f_{i,3} + \epsilon$$

The good news: You know how to estimate the coefficients by linear algebra, optimization, and statistical inference: specifically, linear regression.

The bad news: This model only works for one item, and to be complete, you'd need to build as many models as you have items. Moreover, you're not using other items' information at all to create the model for a given item, so you're not leveraging other pieces of information.

But wait, there's more good news: This solves the "weighting of the features" problem we discussed earlier, because linear regression coefficients *are* weights.

Crap, more bad news: overfitting is *still* a problem, and it comes in the form of having huge coefficients when you don't have enough data (i.e., not enough opinions on given items).

Let's make more rigorous the preceding argument that huge coefficients imply overfitting, or maybe even just a bad model. For example, if two of your variables are exactly the same, or are nearly the same, then the coefficient on one can be 100,000 and the other can be −100,000 and really they add nothing to the model. In general you should always have some *prior* on what a reasonable size would be for your coefficients, which you do by normalizing all of your variables and imagining what an "important" effect would translate to in terms of size of coefficients—anything much larger than that (in an absolute value) would be suspect.

To solve the overfitting problem, you impose a Bayesian prior that these weights shouldn't be too far out of whack—this is done by adding a penalty term for large coefficients. In fact, this ends up being equivalent to adding a prior matrix to the covariance matrix (*http://math babe.org/2013/02/24/the-overburdened-prior/*). That solution depends on a single parameter, which is traditionally called $\lambda$.

But that begs the question: how do you choose $\lambda$? You could do it experimentally: use some data as your training set, evaluate how well you did using particular values of $\lambda$, and adjust. That's kind of what happens in real life, although note that it's not exactly consistent with the idea of estimating what a reasonable size would be for your coefficient.

You can't use this penalty term for large coefficients and assume the "weighting of the features" problem is still solved, because in fact you'd be penalizing some coefficients way more than others if they start out on different scales. The easiest way to get around this is to normalize your variables before entering them into the model, similar to how we did it in Chapter 6. If you have some reason to think certain variables should have larger coefficients, then you can normalize different variables with different means and variances. At the end of the day, the way you normalize is again equivalent to imposing a prior.

A final problem with this prior stuff: although the problem will have a unique solution (as in the penalty will have a unique minimum) if you make $\lambda$ large enough, by that time you may not be solving the problem you care about. Think about it: if you make $\lambda$ absolutely huge, then the coefficients will all go to zero and you'll have no model at all.

## The Dimensionality Problem

OK, so we've tackled the overfitting problem, so now let's think about overdimensionality—i.e., the idea that you might have tens of thousands of items. We typically use both Singular Value Decomposition (SVD) and Principal Component Analysis (PCA) to tackle this, and we'll show you how shortly.

To understand how this works before we dive into the math, let's think about how we reduce dimensions and create "latent features" internally every day. For example, people invent concepts like "coolness," but we can't *directly* measure how cool someone is. Other people exhibit different patterns of behavior, which we internally map or reduce to our one dimension of "coolness." So coolness is an example of a latent feature in that it's unobserved and not measurable directly, and we could think of it as reducing dimensions because perhaps it's a combination of many "features" we've observed about the person and implicitly weighted in our mind.

Two things are happening here: the dimensionality is reduced into a single feature and the latent aspect of that feature.

But in this algorithm, we don't decide which latent factors to care about. Instead we let the machines do the work of figuring out what the important latent features are. "Important" in this context means they explain the variance in the answers to the various questions—in other words, they model the answers efficiently.

Our goal is to build a model that has a representation in a low dimensional subspace that gathers "taste information" to generate recommendations. So we're saying here that taste is *latent* but can be approximated by putting together all the observed information we *do* have about the user.

Also consider that most of the time, the rating questions are binary (yes/no). To deal with this, Hunch created a separate variable for every question. They also found that comparison questions may be better at revealing preferences.

---

## Time to Brush Up on Your Linear Algebra if You Haven't Already

A lot of the rest of this chapter likely won't make sense (and we want it to make sense to you!) if you don't know linear algebra and understand the terminology and geometric interpretation of words like *rank* (hint: the linear algebra definition of that word has nothing to do with ranking algorithms), *orthogonal, transpose, base, span,* and *matrix decomposition*. Thinking about data in matrices as points in space, and what it would mean to transform that space or take subspaces can give you insights into your models, why they're breaking, or how to make your code more efficient. This isn't just a mathematical exercise for the sake of it—although there is elegance and beauty in it—it can be the difference between a start-up that fails and a start-up that gets acquired by eBay. We recommend Khan Academy's excellent free online introduction to linear algebra if you need to brush up your linear algebra skills.

---

## Singular Value Decomposition (SVD)

Hopefully we've given you some intuition about what we're going to do. So let's get into the math now starting with singular value decomposition. Given an $m \times n$ matrix $X$ of rank $k$, it is a theorem from linear algebra that we can always compose it into the product of three matrices as follows:

$$X = USV^\tau$$

where $U$ is $m \times k$, $S$ is $k \times k$, and $V$ is $k \times n$, the columns of $U$ and $V$ are pairwise orthogonal, and $S$ is diagonal. Note the standard statement of SVD is slightly more involved and has U and V both square unitary matrices, and has the middle "diagonal" matrix a rectangular. We'll be using this form, because we're going to be taking approximations to $X$ of increasingly smaller rank. You can find the proof of the existence of this form as a step in the proof of existence of the general form here (*http://goo.gl/GLS6sG*).

Let's apply the preceding matrix decomposition to our situation. $X$ is our original dataset, which has users' ratings of items. We have $m$ users, $n$ items, and $k$ would be the rank of $X$, and consequently would also be an upper bound on the number $d$ of latent variables we decide to care about—note we choose $d$ whereas $m, n$, and $k$ are defined through our training dataset. So just like in k-NN, where $k$ is a tuning parameter (different $k$ entirely—not trying to confuse you!), in this case, $d$ is the tuning parameter.

Each row of $U$ corresponds to a *user*, whereas $V$ has a row for each *item*. The values along the diagonal of the square matrix $S$ are called the "singular values." They measure the importance of each latent variable—the most important latent variable has the biggest singular value.

## Important Properties of SVD

Because the columns of $U$ and $V$ are orthogonal to each other, you can order the columns by singular values via a base change operation. That way, if you put the columns in decreasing order of their corresponding singular values (which you do), then the dimensions are ordered by importance from highest to lowest. You can take lower rank approximation of $X$ by throwing away part of $S$. In other words, replace $S$ by a submatrix taken from the upper-left corner of $S$.

Of course, if you cut off part of $S$ you'd have to simultaneously cut off part of $U$ and part of $V$, but this is OK because you're cutting off the *least important* vectors. This is essentially how you choose the number of latent variables $d$—you no longer have the original matrix $X$ anymore, only an approximation of it, because $d$ is typically much smaller than $k$, but it's still pretty close to $X$. This is what people mean when they talk about "compression," if you've ever heard that term thrown around. There is often an important interpretation to the values in the matrices $U$ and $V$. For example, you can see, by using SVD, that the "most important" latent feature is often something like whether some-one is a man or a woman.

How would you actually use this for recommendation? You'd take $X$, fill in all of the empty cells with the average rating for that item (you don't want to fill it in with 0 because that might mean something in the rating system, and SVD can't handle missing values), and then compute the SVD. Now that you've decomposed it this way, it means that you've captured latent features that you can use to compare users if you want to. But that's not what you want—you want a prediction. If you multiply out the $U$, $S$, and $V^\tau$ together, you get an approximation to $X$—or a prediction, $\hat{X}$—so you can predict a rating by simply look-ing up the entry for the appropriate user/item pair in the matrix $\hat{X}$.

Going back to our original list of issues with nearest neighbors in "Some Problems with Nearest Neighbors" on page 202, we want to avoid the problem of missing data, but that is not fixed by the pre-ceding SVD approach, nor is the computational complexity problem. In fact, SVD is extremely computationally expensive. So let's see how we can improve on that.

## Principal Component Analysis (PCA)

Let's look at another approach for predicting preferences. With this approach, you're still looking for $U$ and $V$ as before, but you don't need $S$ anymore, so you're just searching for $U$ and $V$ such that:

$$X \equiv U \cdot V^\tau$$

Your optimization problem is that you want to minimize the discrep-ency between the actual $X$ and your approximation to $X$ via $U$ and $V$ measured via the squared error:

$$argmin\Sigma_{i,j}\left(x_{i,j} - u_i \cdot v_j\right)^2$$

Here you denote by $u_i$ the row of $U$ corresponding to user $i$, and sim-ilarly you denote by $v_j$ the row of $V$ corresponding to item $j$. As usual, items can include metadata information (so the vector of ages of all the users will be a row in $V$).

Then the dot product $u_i \cdot v_j$ is the *predicted preference* of user $i$ for item $j$, and you want that to be as close as possible to the *actual preference* $x_{i,j}$.

So, you want to find the best choices of $U$ and $V$ that minimize the squared differences between prediction and observation on everything you actually know, and the idea is that if it's really good on stuff you know, it will also be good on stuff you're guessing. This should sound familiar to you—it's mean squared error, like we used for linear regression.

Now you get to choose a parameter, namely the number $d$ defined as *how may latent features you want to use*. The matrix $U$ will have a row for each user and a column for each latent feature, and the matrix $V$ will have a row for each item and a column for each latent feature.

How do you choose $d$? It's typically about 100, because it's more than 20 (as we told you, through the course of developing the product, we found that we had a pretty good grasp on someone if we ask them 20 questions) and it's as much as you care to add before it's computationally too much work.

 The resulting latent features are the basis of a well-defined subspace of the total $n$-dimensional space of potential latent variables. There's no reason to think this solution is unique if there are a bunch of missing values in your "answer" matrix. But that doesn't necessarily matter, because you're just looking for *a solution*.

### Theorem: The resulting latent features will be uncorrelated

We already discussed that correlation was an issue with k-NN, and who wants to have redundant information going into their model? So a nice aspect of these latent features is that they're uncorrelated. Here's a sketch of the proof:

Say we've found matrices $U$ and $V$ with a fixed product $U \cdot V = X$ such that the squared error term is minimized. The next step is to find the best $U$ and $V$ such that their entries are small—actually we're minimizing the sum of the squares of the entries of $U$ and $V$. But we can modify $U$ with any invertible $d \times d$ matrix $G$ as long as we modify $V$ by its inverse: $U \cdot V = (U \cdot G) \cdot (G^{-1} \cdot V) = X$.

Assume for now we only modify with determinant 1 matrices $G$; i.e., we restrict ourselves to volume-preserving transformations. If we ignore for now the size of the entries of $V$ and concentrate only on the size of the entries of $U$, we are minimizing the surface area of a $d$-dimensional parallelepiped in $n$ space (specifically, the one generated by the columns of $U$) where the volume is fixed. This is achieved by making the sides of the parallelepiped mutually orthogonal, which is the same as saying the latent features will be uncorrelated.

But don't forget, we've ignored $V$! However, it turns out that $V$'s rows will also be mutually orthogonal when we force $U$'s columns to be. This is not hard to see if you keep in mind $X$ has its SVD as discussed previously. In fact, the SVD and this form $U \cdot V$ have a lot in common, and some people just call this an SVD algorithm, even though it's not quite.

Now we allow modifications with nontrivial determinant—so, for example, let $G$ be some scaled version of the identity matrix. Then if we do a bit of calculus, it turns out that the best choice of scalar (i.e., to minimize the sum of the squares of the entries of $U$ and of $V$) is in fact the *geometric mean* of those two quantities, which is cool. In other words, we're minimizing the arithmetic mean of them with a single parameter (the scalar) and the answer is the geometric mean.

So that's the proof. Believe us?

## Alternating Least Squares

But how do you do this? How do you actually find $U$ and $V$? In reality, as you will see next, you're not first minimizing the squared error and then minimizing the size of the entries of the matrices $U$ and $V$. You're actually doing both at the same time.

So your goal is to find $U$ and $V$ by solving the optimization problem described earlier. This optimization doesn't have a nice closed formula like ordinary least squares with one set of coefficients. Instead, you need an iterative algorithm like gradient descent. As long as your problem is convex you'll converge OK (i.e., you won't find yourself at a local, but not global, maximum), and you can force your problem to be convex using regularization.

Here's the algorithm:

- Pick a random $V$.

- Optimize $U$ while $V$ is fixed.
- Optimize $V$ while $U$ is fixed.
- Keep doing the preceding two steps until you're not changing very much at all. To be precise, you choose an $\epsilon$ and if your coefficients are each changing by less than $\epsilon$, then you declare your algorithm "converged."

**Theorem with no proof: The preceding algorithm will converge if your prior is large enough**

If you enlarge your prior, you make the optimization easier because you're artificially creating a more convex function—on the other hand, if your prior is huge, then all your coefficients will be zero anyway, so that doesn't really get you anywhere. So actually you might not want to enlarge your prior. Optimizing your prior is philosophically screwed because how is it a *prior* if you're back-fitting it to do what you want it to do? Plus you're mixing metaphors here to some extent by searching for a close approximation of $X$ at the same time you are minimizing coefficients. The more you care about coefficients, the less you care about $X$. But in actuality, you only want to care about $X$.

## Fix V and Update U

The way you do this optimization is user by user. So for user $i$, you want to find:

$$argmin_{u_i} \Sigma_{j \in P_i} \left( p_{i,j} - u_i * v_j \right)^2$$

where $v_j$ is fixed. In other words, you just care about *this user* for now.

But wait a minute, this is the same as linear least squares, and has a closed form solution! In other words, set:

$$u_i = \left( V_{*,i}^\tau V_{*,i} \right)^{-1} V_{*,i}^\tau P_{*i},$$

where $V_{*,i}$ is the subset of $V$ for which you have preferences coming from user $i$. Taking the inverse is easy because it's $d \times d$, which is small. And there aren't that many preferences per user, so solving this many times is really not that hard. Overall you've got a doable update for $U$.

When you fix $U$ and optimize $V$, it's analogous—you only ever have to consider the users that rated that movie, which may be pretty large for popular movies but on average isn't; but even so, you're only ever inverting a $d \times d$ matrix.

Another cool thing: because each user is only dependent on their item's preferences, you can parallelize this update of $U$ or $V$. You can run it on as many different machines as you want to make it fast.

## Last Thoughts on These Algorithms

There are lots of different versions of this approach, but we hope we gave you a flavor for the trade-offs between these methods and how they can be used to make predictions. Sometimes you need to extend a method to make it work in your particular case.

For example, you can add new users, or new data, and keep optimizing $U$ and $V$. You can choose which users you think need more updating to save computing time. Or if they have enough ratings, you can decide not to update the rest of them.

As with any machine learning model, you should perform cross-validation for this model—leave out a bit and see how you did, which we've discussed throughout the book. This is a way of testing overfitting problems.

# Thought Experiment: Filter Bubbles

What are the implications of using error minimization to predict preferences? How does presentation of recommendations affect the feedback collected?

For example, can you end up in local maxima with rich-get-richer effects? In other words, does showing certain items at the beginning give them an unfair advantage over other things? And so do certain things just get popular or not based on luck?

How do you correct for this?

# Exercise: Build Your Own Recommendation System

In Chapter 6, we did some exploratory data analysis on the GetGlue dataset. Now's your opportunity to build a recommendation system with that dataset. The following code isn't for GetGlue, but it is Matt's code to illustrate implementing a recommendation system on a relatively small dataset. Your challenge is to adjust it to work with the GetGlue data.

## Sample Code in Python

```python
import math,numpy

pu     = [[(0,0,1),(0,1,22),(0,2,1),(0,3,1),(0,5,0)],[(1,0,1),
(1,1,32),(1,2,0),(1,3,0),(1,4,1),(1,5,0)],[(2,0,0),(2,1,18),
(2,2,1),(2,3,1),(2,4,0),(2,5,1)],[(3,0,1),(3,1,40),(3,2,1),
(3,3,0),(3,4,0),(3,5,1)],[(4,0,0),(4,1,40),(4,2,0),(4,4,1),
(4,5,0)],[(5,0,0),(5,1,25),(5,2,1),(5,3,1),(5,4,1)]]

pv     = [[(0,0,1),(0,1,1),(0,2,0),(0,3,1),(0,4,0),(0,5,0)],
[(1,0,22),(1,1,32),(1,2,18),(1,3,40),(1,4,40),(1,5,25)],
[(2,0,1),(2,1,0),(2,2,1),(2,3,1),(2,4,0),(2,5,1)],[(3,0,1),
(3,1,0),(3,2,1),(3,3,0),(3,5,1)],[(4,1,1),(4,2,0),(4,3,0),
(4,4,1),(4,5,1)],[(5,0,0),(5,1,0),(5,2,1),(5,3,1),(5,4,0)]]

V = numpy.mat([[ 0.15968384,  0.9441198 ,  0.83651085],
               [ 0.73573009,  0.24906915,  0.85338239],
               [ 0.25605814,  0.6990532 ,  0.50900407],
               [ 0.2405843 ,  0.31848888,  0.60233653],
               [ 0.24237479,  0.15293281,  0.22240255],
               [ 0.03943766,  0.19287528,  0.95094265]])

print V

U = numpy.mat(numpy.zeros([6,3]))
L = 0.03

for iter in xrange(5):

    print "\n----- ITER %s -----"%(iter+1)

    print "U"
    urs = []
    for uset in pu:
        vo = []
        pvo = []
```

```
        for i,j,p in uset:
            vor = []
            for k in xrange(3):
                vor.append(V[j,k])
            vo.append(vor)
            pvo.append(p)
        vo = numpy.mat(vo)
        ur = numpy.linalg.inv(vo.T*vo +
            L*numpy.mat(numpy.eye(3))) *
            vo.T * numpy.mat(pvo).T
        urs.append(ur.T)
    U = numpy.vstack(urs)
    print U

    print "V"
    vrs = []
    for vset in pv:
        uo = []
        puo = []
        for j,i,p in vset:
            uor = []
            for k in xrange(3):
                uor.append(U[i,k])
            uo.append(uor)
            puo.append(p)
        uo = numpy.mat(uo)
            vr = numpy.linalg.inv(uo.T*uo  +  L*numpy.mat(num
py.eye(3))) * uo.T * numpy.mat(puo).T
        vrs.append(vr.T)
    V = numpy.vstack(vrs)
    print V

    err = 0.
    n = 0.
    for uset in pu:
        for i,j,p in uset:
            err += (p - (U[i]*V[j].T)[0,0])**2
            n += 1
    print math.sqrt(err/n)

print
print U*V.T
```

# CHAPTER 9

# Data Visualization and Fraud Detection

There are two contributors for this chapter, Mark Hansen, a professor at Columbia University, and Ian Wong, an inference scientist at Square. (That's where he was in November 2012 when he came to the class. He now works at Prismatic.) These two speakers and sections don't share a single cohesive theme between them, although both will discuss data visualization...and both have lived in California! More seriously, both are thoughtful people (like all our contributors) who have thought deeply about themes and questions such as what makes good code, the nature of programming languages as a form of expression, and the central question of this book: what is data science?

## Data Visualization History

First up is Mark Hansen, who recently came from UCLA via a sabbatical at the *New York Times* R & D Lab to Columbia University with a joint appointment in journalism and statistics, where he heads the Brown Institute for Media Innovation. He has a PhD in statistics from Berkeley, and worked at Bell Labs (there's Bell Labs again!) for several years prior to his appointment at UCLA, where he held an appointment in statistics, with courtesy appointments in electrical engineering and design/media art. He is a renowned data visualization expert and also an energetic and generous speaker. We were lucky to have him on a night where he'd been drinking an XXL latte from Starbucks (we refuse to use their made-up terminology) to highlight his natural effervescence.

Mark will walk us through a series of influences and provide historical context for his data visualization projects, which he will tell us more about at the end. Mark's projects are genuine works of art—installations appearing in museums and public spaces. Rachel invited him because his work and philosophy is inspiring and something to aspire to. He has set his own course, defined his own field, exploded boundaries, and constantly challenges the status quo. He's been doing data visualization since before data visualization was cool, or to put it another way, we consider him to be one of the fathers of data visualization. For the practical purposes of becoming better at data visualization yourself, we'll give you some ideas and directions at the end of the chapter.

## Gabriel Tarde

Mark started by telling us a bit about Gabriel Tarde, who was a sociologist who believed that the social sciences had the capacity to produce vastly more data than the physical sciences.

As Tarde saw it, the physical sciences *observe from a distance*: they typically model or incorporate models that talk about an aggregate in some way—for example, a biologist might talk about the function of the aggregate of our cells. What Tarde pointed out was that this is a deficiency; it's basically brought on by a lack of information. According to Tarde, we should instead be tracking every cell.

In the social realm we can do the analog of this, if we replace cells with people. We can collect a huge amount of information about individuals, especially if they offer it up themselves through Facebook.

But wait, are we not missing the forest for the trees when we do this? In other words, if we focus on the microlevel, we might miss the larger cultural significance of social interaction. Bruno Latour, a contemporary French sociologist, weighs in with his take on Tarde in Tarde's Idea of Quantification:

> But the *whole* is now nothing more than a provisional visualization which can be modified and reversed at will, by moving back to the individual components, and then looking for yet other tools to regroup the same elements into alternative assemblages.
>
> — Bruno Latour

In 1903, Tarde even foresaw the emergence of Facebook, as a sort of "daily press":

If statistics continues to progress as it has done for several years, if the information which it gives us continues to gain in accuracy, in dispatch, in bulk, and in regularity, a time may come when **upon the accomplishment of every social event a figure will at once issue forth automatically, so to speak, to take its place on the statistical registers that will be continuously communicated to the public and spread abroad pictorially by the daily press**. Then, at every step, at every glance cast upon poster or newspaper, we shall be assailed, as it were, with statistical facts, with **precise and condensed knowledge of all the peculiarities of actual social conditions, of commercial gains or losses, of the rise or falling off of certain political parties, of the progress or decay of a certain doctrine, etc.**, in exactly the same way as we are assailed when we open our eyes by the vibrations of the ether which tell us of the approach or withdrawal of such and such a so-called body and of many other things of a similar nature.

— Tarde

Mark then laid down the theme of his lecture:

Change the instruments and you will change the entire social theory that goes with them.

— Bruno Latour

Kind of like that famous physics cat, Mark (and Tarde) want us to newly consider both the way the structure of society changes as we observe it, and ways of thinking about the relationship of the individual to the aggregate.

In other words, the past nature of data collection methods forced one to consider aggregate statistics that one can reasonably estimate by subsample—means, for example. But now that one can actually get one's hands on all data and work with all data, one no longer should focus only on the kinds of statistics that make sense in the aggregate, but also one's own individually designed statistics—say, coming from graph-like interactions—that are now possible due to finer control. Don't let the dogma that resulted from past restrictions guide your thinking when those restrictions no longer hold.

## Mark's Thought Experiment

As data become more personal, as we collect more data about individuals, what new methods or tools do we need to express the fundamental relationship between ourselves and our communities, our communities and our country, our country and the world?

Could we ever be satisfied with poll results or presidential approval ratings when we can see the complete trajectory of public opinions, both individuated and interacting?

To which we add: would we actually want to live in a culture where such information is so finely tracked and available?

## What Is Data Science, Redux?

Mark reexamined the question that Rachel posed and attempted to answer in the first chapter, as he is keen on reexamining everything. He started the conversation with this undated quote from our own John Tukey:

> The best thing about being a statistician is that you get to play in everyone's backyard.
>
> — John Tukey

Let's think about that again—is it so great? Is it even reasonable? In some sense, to think of us as playing in *other* people's yards, with *their* toys, is to draw a line between "traditional data fields" and "everything else."

It's maybe even implying that all our magic comes from the traditional data fields (math, stats, CS), and we're some kind of super humans because we're uber-nerds. That's a convenient way to look at it from the perspective of our egos, of course, but it's perhaps too narrow and arrogant.

And it begs the question: what is "traditional" and what is "everything else," anyway?

In Mark's opinion, "everything else" should include fields from social science and physical science to education, design, journalism, and media art. There's more to our practice than being technologists, and we need to realize that technology itself emerges out of the natural needs of a discipline. For example, geographic information systems (GIS) emerged from geographers, and text data mining emerged from digital humanities.

In other words, it's not math people ruling the world, but rather domain practices being informed by techniques growing organically from those fields. When data intersects their practice, each practice is learning differently; their concerns are unique to that practice.

Responsible data science integrates those lessons, and it's not a purely mathematical integration. It could be a way of describing events, for example. Specifically, we're saying that it's not necessarily a quantifiable thing.

Bottom-line: it's possible that the language of data science has *something* to do with social science just as it has *something* to do with math.

You might not be surprised to hear that, when Mark told us about his profile as a data scientist, the term he coined was "expansionist."

## Processing

Mark then described the programming language called Processing (*http://processing.org*) in the context of a programming class he gave to artists and designers. He used it as an example of what is different when a designer, versus an engineer, takes up looking at data or starts to code. A good language is inherently structured or designed to be expressive of the desired tasks and ways of thinking of the people using it.

One approach to understanding this difference is by way of another thought experiment. Namely, what is the use case for a language for artists? Contrast this with what a language such as R needs to capture for statisticians or data scientists (randomness, distributions, vectors, and data, for example).

In a language for artists, you'd want to be able to specify shapes, to faithfully render whatever visual thing you had in mind, to sketch, possibly in 3D, to animate, to interact, and most importantly, to *publish*.

Processing is Java-based, with a simple "publish" button, for example. The language is adapted to the practice of artists. Mark mentioned that teaching designers to code meant, for him, stepping back and talking about iteration, if statements, and so on—in other words, stuff that seemed obvious to him but is not obvious to someone who is an artist. He needed to unpack his assumptions, which is what's fun about teaching to the uninitiated.

## Franco Moretti

Mark moved on to discussing close versus distant reading of texts. He mentioned Franco Moretti, a literary scholar from Stanford.

Franco thinks about "distant reading," which means trying to get a sense of what someone's talking about without reading line by line. This leads to PCA-esque thinking, a kind of dimension reduction of novels (recall we studied dimension reduction techniques in Chapter 8).

Mark holds this up as a cool example of how ideally data science integrates the ways that experts in various fields already figure stuff out. In other words, we don't just go into their backyards and play; maybe instead we go in and watch *them* play, and then formalize and inform their process with our own bells and whistles. In this way they can teach us new games, games that actually expand our fundamental conceptions of data and the approaches we need to analyze them.

# A Sample of Data Visualization Projects

Here are some of Mark's favorite visualization projects, and for each one he asks us: is this your idea of data visualization? What's data?

Figure 9-1 is a projection onto a power plant's steam cloud. The size of the green projection corresponds to the amount of energy the city is using.

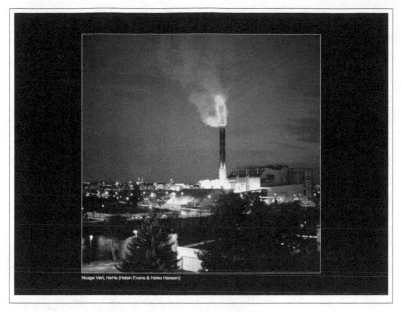

Nuage Vert, HeHe (Helen Evans & Heiko Hansen)

*Figure 9-1. Nuage Vert by Helen Evans and Heiko Hansen (http://
youtu.be/l_4rTQCWItw)*

In *One Tree* (Figure 9-2) the artist cloned trees and planted the ge-
netically identical seeds in several areas. It displays, among other
things, the environmental conditions in each area where they are
planted.

*Figure 9-2. One Tree by Natalie Jeremijenko (http://boingboing.net/2003/05/16/natalie-jeremijenkos.html)*

Figure 9-3 shows *Dusty Relief*, in which the building collects pollution around it, displayed as dust.

Dusty Relief, New Territories

*Figure 9-3. Dusty Relief from New Territories (http://www.new-territories.com/roche2002bis.htm)*

*Project Reveal* (Figure 9-4) is a kind of magic mirror that wirelessly connects using facial recognition technology and gives you information about yourself. According to Mark, as you stand at the mirror in the morning, you get that "come-to-Jesus moment."

*Figure 9-4. Project Reveal from the New York Times R & D lab (http://nytlabs.com/projects/mirror.html)*

The SIDL is headed by Laura Kurgan, and in this piece shown in Figure 9-5, she flipped Google's crime statistics. She went into the prison population data, and for every incarcerated person, she looked at their home address, measuring per home how much money the state was spending to keep the people who lived there in prison. She discovered that some blocks were spending $1,000,000 to keep people in prison. The moral of this project is: just because you can put something on the map, doesn't mean you should. It doesn't mean there's a new story. Sometimes you need to dig deeper and flip it over to get a new story.

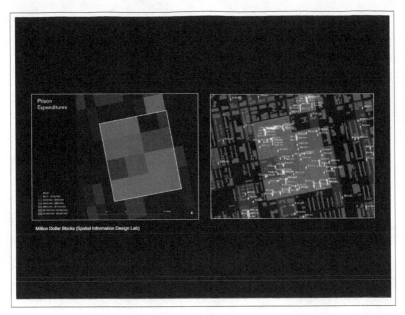

*Figure 9-5. Million Dollar Blocks from Spatial Information Design Lab (SIDL) (http://www.spatialinformationdesignlab.org)*

# Mark's Data Visualization Projects

Now that we know some of Mark's influences and philosophy, let's look at some of his projects to see how he puts them into practice.

## New York Times Lobby: Moveable Type

Mark walked us through a project he did with Ben Rubin—a media artist and Mark's collaborator of many years—for the *New York Times* on commission. (Mark later went to the *New York Times* R & D Lab on sabbatical.) Figure 9-6 shows it installed in the lobby of the *Times'* midtown Manhattan headquarters at 8th Avenue and 42nd Street.

*Figure 9-6. Moveable Type, the New York Times lobby, by Ben Rubin and Mark Hansen*

It consists of 560 text displays—two walls with 280 displays on each—and they cycle through various scenes that each have a theme and an underlying data science model.

In one there are waves upon waves of digital ticker-tape–like scenes that leave behind clusters of text, and where each cluster represents a different story from the paper. The text for a given story highlights phrases that make a given story different from others in an information-theory sense.

In another scene, the numbers coming out of stories are highlighted, so you might see "18 gorillas" on a given display. In a third scene, crossword puzzles play themselves accompanied by sounds of pencils writing on paper.

Figure 9-7 shows an example of a display box, which are designed to convey a retro vibe. Each box has an embedded Linux processor running Python, and a sound card that makes various sounds—clicking, typing, waves—depending on what scene is playing.

*Figure 9-7. Display box for Moveable Type*

The data is collected via text from *New York Times* articles, blogs, and search engine activity. Every sentence is parsed using Stanford natural language processing techniques (*http://nlp.stanford.edu*), which diagram sentences.

Altogether there are about 15 scenes so far, and it's written in code so one can keep adding to it. Here's a YouTube interview (*http://goo.gl/ yhCG69*) with Mark and Ben about the exhibit.

## Project Cascade: Lives on a Screen

Mark next told us about Cascade, which was a joint work with Jer Thorp—data artist-in-residence at the *New York Times* —in partnership with bit.ly. Cascade came about from thinking about how people share *New York Times* links on Twitter.

The idea was to collect enough data so that you could see people browse, encode the link in bit.ly, tweet that encoded link, see other people click on that tweet, watch bit.ly decode the link, and then see those people browse the *New York Times*. Figure 9-8 shows the visualization of that entire process, much like Tarde suggested we should do.

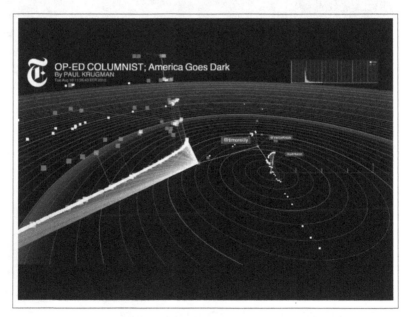

*Figure 9-8. Project Cascade by Jer Thorp and Mark Hansen*

There were of course data decisions to be made: a loose matching of tweets and clicks through time, for example. If 17 different tweets provided the same URL, they couldn't know which tweet/link someone clicked on, so they guessed (the guess actually involves probabilistic matching on timestamps so at least it's an educated guess). They used the Twitter map of who follows who—if someone you follow tweets about something before you do, then it counts as a retweet.

Here's a video (*http://goo.gl/uAOcg8*) about Project Cascade from the *New York Times*.

This was done two years ago, and Twitter has gotten a lot bigger since then.

## Cronkite Plaza

Next Mark told us about something he was working on with both Jer and Ben. It's also news-related, but it entailed projecting something on the outside of a building rather than in the lobby; specifically, the communications building at UT Austin, in Cronkite Plaza, pictured in Figure 9-9.

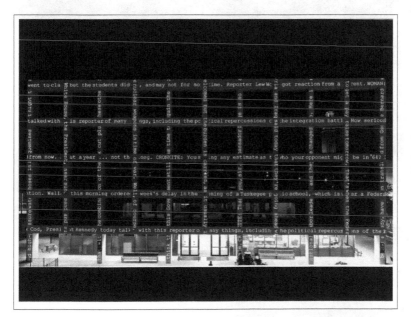

*Figure 9-9. And That's The Way It Is, by Jer Thorp, Mark Hansen, and Ben Rubin*

It's visible every evening at Cronkite Plaza, with scenes projected onto the building via six different projectors. The majority of the projected text is sourced from Walter Cronkite's news broadcasts, but they also used local closed-captioned news sources. One scene extracted the questions asked during local news—things like "How did she react?" or "What type of dog would you get?"

## eBay Transactions and Books

Again working jointly with Jer Thorp, Mark investigated a day's worth of eBay's transactions that went through Paypal and, for whatever reason, two years of book sales. How do you visualize this? Take a look at their data art-visualization-installation commissioned by eBay for the 2012 ZERO1 Biennial in Figure 9-10 and at the yummy underlying data in Figure 9-11.

*Figure 9-10. Before Us is the Salesman's House (2012), by Jer Thorp and Mark Hansen*

```
4106.00,CHINESE OLD SIGNED CERAMIC SHUDEI BONSAI POT KYUSU NR
712.50,Omron HEM-650 Wrist Blood Pressure Monitor with APS
499.99,Panasonic TC-P42X3 42" Viera Plasma HDTV
491.00,$500 Best Buy Gift Card for $491.00
372.23,Droid Incredible 2 (Verizon)
346.99,Hondo Chiquita Guitar
329.70,IMATION CD-RW 48 SPINDLES
324.25,1979 Camellias of Yunnan Souvenir Miniature Stamp sheet
287.00,DIESEL New $690 Leather Jacket Coat M
279.79,NEW CALLAWAY TOP FLITE COMPLETE MRH GOLF CLUBS SET BLUE
275.88,AMAZING LABRADORITE 925 SILVER CUFF BRACELET BA31
227.16,NEW Chefs Banquet Survival Emergency Food 330 servings
212.30,Authentic Prada Blue Bag
203.50,Superb Chinese Carved Jade Plaque 18th C.
189.00,NEW IDF Tactical Combat Vest with Removable Backpack
187.98,HP OfficeJet Pro 8500A PLUS A1O Printer    A910g      NEW
182.90,Cisco Small Business RV042 Dual WAN VPN Router -   RV042
174.95,PHILIPPE STARCK Black VEILED Stainless VEIL NEW! PH5022
173.23,14K. GOLD CHANDELIERS EARRINGS NATURAL BLUE TOPAZ GEMS
137.99,FUJI FINEPIX JZ300 10x 12MP DIGITAL CAMERA SILVER NEW
126.00,New Stock White/Ivory Wedding Dress Sz:6/8/10/12/14/16
124.64,Lot of 700 HP 02 Mixed Colors Empty Ink Tank Cartridges
116.40,Goddess of Wisdom 2001 Barbie Doll NIB Mint Condition
112.08,Littmann Cardiology III Stethescope
111.11,ALLEN EDMONDS PARK AVENUE OX BLACK 8 D MEDIUM $325
105.95,REPRODUCTION GERMAN WWII WOOL SOCKS SIZE 2 RING (9-10)
99.00,TOSHIBA SATELLITE L305-S5875 15.4" WXGA  LCD SCREEN
97.58,calvin klein men's slim fit plaid suit pants
90.00,Pear Shape Diamond
80.30,HAPPY CHLOE DUCK 2010 SWAROVSKI  RETIRED  #1041293
79.99,Cisco 2600 series 2611 2-port Ethernet Router CCNA CCNP
77.77,PRO BICYCLE MECHANICS XLC TOOL KIT 33pc BIKE REPAIR SET
74.90,NEW LG VX10000 VOYAGER QWERTY TXT MP3 PHONE VERIZON
69.94,Modern Jacquard Bedding Comforter Set Queen Black/Grey
67.49,pearl gameboy advanced SP metal case 18 gamesECT.
```

*Figure 9-11. The underlying data for the eBay installation*

Here's their ingenious approach: They started with the text of *Death of a Salesman* by Arthur Miller. They used a mechanical turk mechanism (we discussed what this means in Chapter 7) to locate objects in the text that you can buy on eBay.

When an object is found it moves it to a special bin, e.g., "chair" or "flute" or "table." When it has a few collected buyable objects, it then takes the objects and sees where they are all for sale on the day's worth of transactions, and looks at details on outliers and such. After examining the sales, the code will find a zip code in some quiet place like Montana.

Then it flips over to the book sales data, looks at all the books bought or sold in that zip code, picks a book (which is also on Project Gutenberg), and begins to read *that* book and collect "buyable" objects from it. And it keeps going. Here's a video showing the process (*https://vimeo.com/50146828*).

## Public Theater Shakespeare Machine

The last piece Mark showed us is a joint work with Rubin and Thorp, installed in the lobby of the Public Theater, shown in Figure 9-12. The piece is an oval structure with 37 bladed LED displays installed above the theater's bar.

There's one blade for each of Shakespeare's plays. Longer plays are in the long end of the oval—you see *Hamlet* when you come in.

*Figure 9-12. Shakespeare Machine, by Mark, Jer, and Ben*

The data input is the text of each play. Each scene does something different—for example, it might collect noun phrases that have something to do with the body from each play, so the "Hamlet" blade will only show a body phrase from *Hamlet*. In another scene, various kinds of combinations or linguistic constructs are mined, such as three-word phrases like "high and might" or "good and gracious" or compound-word phrases like "devilish-holy," "heart-sore," or "hard-favoured."

Note here that the digital humanities, through the MONK Project (*http://monkproject.org*), offered intense XML descriptions of the plays. Every single word is given hooha, and there's something on the order of 150 different parts of speech.

As Mark said, it's Shakespeare, so it stays awesome no matter what you do. But they're also successively considering words as symbols, or as thematic, or as parts of speech.

So then let's revisit the question Mark asked before showing us all these visualizations: what's data? It's *all* data.

Here's one last piece of advice from Mark on how one acquires data. Be a good investigator: a small polite voice which asks for data usually gets it.

## Goals of These Exhibits

These exhibits are meant to be graceful and artistic, but they should also teach something or tell a story. At the same time, we don't want to be overly didactic. The aim is to exist in between art and information. It's a funny place: increasingly we see a flattening effect when tools are digitized and made available, so that statisticians can code like a designer—we can make things that look like design, but is it truly design—and similarly designers can make something that looks like data or statistics, but is it really?

# Data Science and Risk

Next we had a visitor from San Francisco—Ian Wong, who came to tell us about doing data science on the topic of risk. Ian is an inference scientist at Square, and he previously dropped out of the electrical engineering PhD program at Stanford where he did research in machine learning. (He picked up a couple master's degrees in statistics and electrical engineering along the way.) Since coming to speak to the class, he left Square and now works at Prismatic, a customized newsfeed.

Ian started with three takeaways:

*Doing machine learning != writing R scripts*
Machine learning (ML) is founded in math, expressed in code, and assembled into software. You need to develop good software engineering practices, and learn to write readable and reusable code. Write code for the reader and not the writer, as production code is reread and built upon many more times by other people than by you.

*Data visualization != producing a nice plot*
Visualizations should be pervasive and embedded in the environment of a good company. They're integrated in products and processes. They should enable action.

*ML and data visualization together augment human intelligence*
We have limited cognitive abilities as human beings. But by leveraging data, we can build ourselves exoskeletons that enable us to comprehend and navigate the information world.

## About Square

Square was founded in 2009 by Jack Dorsey and Jim McKelvey. The company grew from 50 employees in 2011 to over 500 in 2013.

The mission of the company is to make commerce easy for everyone. As Square's founders see it, transactions are needlessly complicated. It takes too much for a merchant to figure out how to accept payments. For that matter, it's too complicated for buyers to pay as well. The question they set out to answer is "how do we make transactions simple and easy?"

Here's how they do it. Merchants can sign up with Square, download the Square Register app, and receive a credit card reader in the mail. They can then plug the reader into the phone, open the app, and take payments. The little plastic square enables small merchants (any size really) to accept credit card transactions. Local hipster coffee shops seem to have been early adopters if Portland and San Francisco are any indication. On the consumer's side, they don't have to do anything special, just hand over their credit cards. They won't experience anything unusual, although they do sign on the iPad rather than on a slip of paper.

It's even possible to buy things hands-free using the Square. When the buyer chooses to pay through Square Wallet on their phones, the buyer's name will appear on the merchant's Register app and all the merchant has to do is to tap on the name.

Square wants to make it easy for sellers to sign up for their service and to accept payments. Of course, it's also possible that somebody may sign up and try to abuse the service. They are, therefore, very careful at Square to avoid losing money on sellers with fraudulent intentions or bad business models.

# The Risk Challenge

In building a frictionless experience for buyers and sellers, Square also has to watch out for the subset of users who abuse the service. Suspicious or unwanted activity, such as fraud, not only undermines customer trust, but is illegal and impacts the company's bottom line. So creating a robust and highly efficient risk management system is core to the payment company's growth.

But how does Square detect bad behavior efficiently? Ian explained that they do this by investing in machine learning with a healthy dose of visualization.

### Detecting suspicious activity using machine learning

Let's start by asking: what's suspicious? If we see lots of micro transactions occurring, say, or if we see a sudden, high frequency of transactions, or an inconsistent frequency of transactions, that might raise our eyebrows.

Here's an example. Say John has a food truck, and a few weeks after he opens, he starts to pass $1,000 transactions through Square. (One possibility: John might be the kind of idiot that puts gold leaf on hamburgers.) On the one hand, if we let money go through, Square is on the spot in case it's a bad charge. Technically the fraudster—who in this case is probably John—would be liable, but our experience is that usually fraudsters are insolvent, so it ends up on Square to foot the bill.

On the other hand, if Square stops payment on what turns out to be a real payment, it's bad customer service. After all, what if John is innocent and Square denies the charge? He will probably be pissed at Square —and he may even try to publicly sully Square's reputation—but in any case, the trust is lost with him after that.

This example crystallizes the important challenges Square faces: false positives erode customer trust, false negatives cost Square money.

To be clear, there are actually *two* kinds of fraud to worry about: seller-side fraud and buyer-side fraud. For the purpose of this discussion, we'll focus on seller-side fraud, as was the case in the story with John.

Because Square processes millions of dollars worth of sales per day, they need to gauge the plausibility of charges systematically and automatically. They need to assess the risk level of every event and entity in the system.

So what do they do? Before diving in, Ian sketched out part of their data schema, shown in Figures 9-13, 9-14, and 9-15.

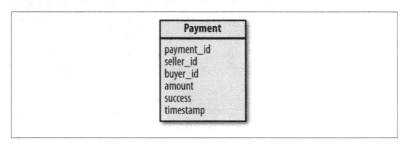

*Figure 9-13. Payment schema*

*Figure 9-14. Seller schema*

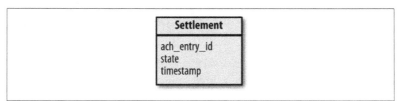

*Figure 9-15. Settlement schema*

There are three types of data represented here:

- Payment data, where we can assume the fields are transaction_id, seller_id, buyer_id, amount, success (0 or 1), and timestamp.
- Seller data, where we can assume the fields are seller_id, sign_up_date, business_name, business_type, and business_location.
- Settlement data, where we can assume the fields are settlement_id, state, and timestamp.

It's important to note that Square settles with its customers a full day after the initial transaction, so their process doesn't have to make a decision within microseconds. They'd like to do it quickly of course, but in certain cases, there is time for a phone call to check on things.

Here's the process shown in Figure 9-16: given a bunch (as in millions) of payment events and their associated date (as shown in the data schema earlier), they throw each through the risk models, and then send some iffy-looking ones on to a "manual review." An ops team will then review the cases on an individual basis. Specifically, anything that looks rejectable gets sent to ops, who follow up with the merchants. All approved transactions are settled (yielding an entry in the `settle ment` table).

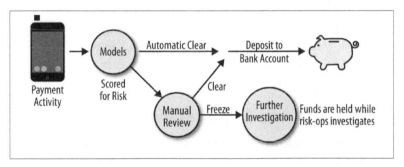

*Figure 9-16. Risk engine*

Given the preceding process, let's focus on how they set up the risk models. You can think of the model as a function from payments to labels (e.g., good or bad). Putting it that way, it kind of sounds like a straightforward supervised learning problem. And although this problem shares *some* properties with that, it's certainly not that simple —they don't reject a payment and then merely stand pat with that label, because, as we discussed, they send it on to an ops team to assess it independently. So in actuality they have a pretty complicated set of labels, including when a charge is initially rejected but later decide it's OK, or it's initially accepted but on further consideration might have been bad, or it's *confirmed* to have been bad, or confirmed to have been OK, and the list goes on.

Technically we would call this a *semi-supervised learning problem*, straddling the worlds of supervised and unsupervised learning. But it's useful to note that the "label churn" settles down after a few months when the vast majority of chargebacks have been received, so they could treat the problem as strictly supervised learning if you go far enough back in time. So while they can't trust the labels on recent data, for the purpose of this discussion, Ian will describe the easier case of solving the supervised part.

Now that we've set the stage for the problem, Ian moved on to describing the *supervised learning recipe* as typically taught in school:

- Get data.
- Derive features.
- Train model.
- Estimate performance.
- Publish model!

But transferring this recipe to the real-world setting is not so simple. In fact, it's not even clear that the order is correct. Ian advocates thinking about the *objective* first and foremost, which means bringing performance estimation to the top of the list.

## The Trouble with Performance Estimation

So let's do that: focus on performance estimation. Right away Ian identifies three areas where we can run into trouble.

### Defining the error metric

How do we measure whether our learning problem is being modeled well? Let's remind ourselves of the various possibilities using the truth table in Table 9-1.

*Table 9-1. Actual versus predicted table, also called the Confusion Matrix*

|  | Actual = True | Actual = False |
| --- | --- | --- |
| Predicted = True | *TP* (true positive) | *FP* (false positive) |
| Predicted = False | *FN* (false negative) | *TN* (true negative) |

The most straightforward performance metric is *Accuracy*, which is defined using the preceding notation as the ratio:

$$Accuracy = \frac{TP + TN}{TP + TN + FP + FN}$$

Another way of thinking about accuracy is that it's the probability that your model gets the right answer. Given that there are very few positive examples of fraud—at least compared with the overall number of transactions—accuracy is not a good metric of success, because the "everything looks good" model, or equivalently the "nothing looks fraudulent" model, is dumb but has good accuracy.

Instead, we can estimate performance using *Precision* and *Recall*. Precision is defined as the ratio:

$$Precision = \frac{TP}{TP + FP}$$

or the probability that a transaction branded "fraudulent" is actually fraudulent.

Recall is defined as the ratio:

$$Recall = \frac{TP}{TP + FN}$$

or the probability that a truly fraudulent transaction is caught by the model.

The decision of which of these metrics to optimize for depends on the costs of uncaught bad transactions, which are easy to measure, versus overly zealous caught transactions, which are much harder to measure.

## Defining the labels

Labels are what Ian considered to be the "neglected" half of the data. In undergrad statistics education and in data mining competitions, the availability of labels is often taken for granted. But in reality, labels are tough to define and capture, while at the same time they are vitally important. It's not related to just the objective function; it *is* the objective.

In Square's setting, defining the label means being precise about:

- What counts as a suspicious activity?
- What is the right level of granularity? An event or an entity (or both)?
- Can we capture the label reliably? What other systems do we need to integrate with to get this data?

Lastly, Ian briefly mentioned that label noise can acutely affect prediction problems with *high class imbalance* (e.g., very few positive samples).

## Challenges in features and learning

Ian says that features codify your domain knowledge. Once a machine learning pipeline is up and running, most of the modeling energy should be spent trying to figure out better ways to describe the domain (i.e., coming up with new features). But you have to be aware of when these features can actually be learned.

More precisely, when you are faced with a class imbalanced problem, you have to be careful about overfitting. The sample size required to learn a feature is proportional to the population of interest (which, in this case, is the "fraud" class).

For example, it can get tricky dealing with categorical variables with many levels. While you may have a zip code for every seller, you don't have enough information in knowing the zip code alone because so few fraudulant sellers share zip codes. In this case, you want to do some clever binning of the zip codes. In some cases, Ian and his team create a submodel within a model just to reduce the dimension of certain features.

There's a second data sparsity issue, which is the *cold start* problem with new sellers. You don't know the same information for all of your sellers, especially for new sellers. But if you are too conservative, you risk starting off on the wrong foot with new customers.

Finally, and this is typical for predictive algorithms, you need to tweak your algorithm to fit the problem setting, which in this case is akin to finding a needle in a haystack. For example, you need to consider whether features interact linearly or nonlinearly, and how to adjust model training to account for class imbalance: should you adjust the weights for each class? How about the sampling scheme in an ensemble learner?

You also have to be aware of adversarial behavior, which is a way of saying that someone is actually scheming against you. Here's an example from ecommerce: if a malicious buyer figures out that you are doing fraud detection by address resolution via exact string matching, then he can simply sign up with 10 new accounts, each with a slight misspelling of same address. You now need to know how to resolve these variants of addresses, and anticipate the escalation of adverserial behavior. Because models degrade over time as people learn to game them, you need to continually track performance and retrain your models.

## What's the Label?

Here's another example of the trickiness of labels. At DataEDGE, a conference held annually at UC Berkeley's School of Information, in a conversation between Michael Chui of the McKinsey Global Institute, and Itamar Rosenn—Facebook's first data scientist (hired in 2007)—Itamar described the difficulties in defining an "engaged user." What's the definition of *engaged*? If you want to predict whether or not a user is engaged, then you need some notion of engaged if you're going to label users as engaged or not. There is no one obvious definition, and, in fact, a multitude of definitions might work depending on the context—there is no ground truth! Some definitions of engagement could depend on the frequency or rhythm with which a user comes to a site, or how much they create or consume content. It's a semi-supervised learning problem where you're simultaneously trying to define the labels as well as predict them.

# Model Building Tips

Here are a few good guidelines to building good production models:

*Models are not black boxes*
> You can't build a good model by assuming that the algorithm will take care of everything. For instance, you need to know why you are misclassifying certain people, so you'll need to roll up your sleeves and dig into your model to look at what happened. You essentially form a narrative around the mistakes.

*Develop the ability to perform rapid model iterations*
> Think of this like experiments you'd do in a science lab. If you're not sure whether to try A or B, then try both. Of course there's a limit to this, but most of the time people err on the "not doing it enough" side of things.

*Models and packages are not a magic potion*
> When you hear someone say, "So which models or packages do you use?" then you've run into someone who doesn't get it.

Ian notices that the starting point of a lot of machine learning discussions revolves around what algorithm or package people use. For instance, if you're in R, people get caught up on whether you're using `randomForst`, `gbm`, `glmnet`, `caret`, `ggplot2`, or `rocr`; or in `scikit-learn` (Python), whether you're using the `RandomForestClassifier` or `RidgeClassifier`. But that's losing sight of the forest.

## Code readability and reusability

So if it's not about the models, what is it really about then? It's about your ability to use and reuse these packages, to be able to swap any and all of these models with ease. Ian encourages people to concentrate on readability, reusability, correctness, structure, and hygiene of these code bases.

And if you ever dig deep and implement an algorithm yourself, try your best to produce understandable and extendable code. If you're coding a random forest algorithm and you've hardcoded the number of trees, you're backing yourself (and anyone else who uses that algorithm) into a corner. Put a parameter there so people can reuse it. Make it tweakable. Favor composition. And write tests, for pity's sake. Clean code and clarity of thought go together.

At Square they try to maintain reusability and readability by structuring code in different folders with distinct, reusable components that provide semantics around the different parts of building a machine learning model:

*Model*
> The learning algorithms

*Signal*
> Data ingestion and feature computation

*Error*
> Performance estimation

*Experiment*
> Scripts for exploratory data analysis and experiments

*Test*
> Test all the things

They only write scripts in the experiments folder where they either tie together components from model, signal, and error, or conduct exploratory data analysis. Each time they write a script, it's more than just a piece of code waiting to rot. It's an experiment that is revisited over and over again to generate insight.

What does such a discipline give you? Every time you run an experiment, you should incrementally increase your knowledge. If that's not happening, the experiment is not useful. This discipline helps you make sure you don't do the same work again. Without it you can't even figure out the things you or someone else has already attempted. Ian further claims that "If you don't write production code, then you're not productive."

For more on what every project directory should contain, see Project Template (*http://projecttemplate.net/*) by John Myles White. For those students who are using R for their classes, Ian suggests exploring and actively reading GitHub's repository of R code. He says to try writing your own R package, and make sure to read Hadley Wickham's devtools wiki (*http://adv-r.had.co.nz/*). Also, he says that developing an aesthetic sense for code is analogous to acquiring the taste for beautiful proofs; it's done through rigorous practice and feedback from peers and mentors.

For extra credit, Ian suggests that you contrast the implementations of the caret package (*http://cran.r-project.org/web/packages/caret/*

*index.html*) with scikit-learn (*https://github.com/scikit-learn/scikit-learn*). Which one is more extendable and reusable? Why?

### Get a pair!

Learning how to code well is challenging. And it's even harder to go about it alone. Imagine trying to learn Spanish (or your favorite language) and not being able to practice with another person.

Find a partner or a team who will be willing to pair program and conduct rigorous code reviews. At Square, every single piece of code is reviewed by at least another pair of eyes. This is an important practice not only for error checking purposes, but also to ensure shared ownership of the code and a high standard of code quality.

Here's something to try once you find a programming buddy. Identify a common problem. Set up a workstation with one monitor and two sets of keyboard and mouse. Think of it as teaming up to solve the problem: you'll first discuss the overall strategy of solving the problem, and then move on to actually implementing the solution. The two of you will take turns being the "driver" or the "observer." The driver writes the code, while the observer reviews the code and strategizes the plan of attack. While the driver is busy typing, the observer should constantly be asking "do I understand this code?" and "how can this code be clearer?" When confusion arises, take the time to figure out the misunderstandings (or even lack of understanding) together. Be open to learn and to teach. You'll pick up nuggets of knowledge quickly, from editor shortcuts to coherent code organization.

The driver and observer roles should switch periodically throughout the day. If done right, you'll feel exhausted after several hours. Practice to improve pairing endurance.

And when you don't get to pair program, develop the habit to check in code using Git. Learn about Git workflows, and give each other constructive critiques on pull requests. Think of it as peer review in academia.

### Productionizing machine learning models

Here are some of the toughest problems in doing real-world machine learning:

1. How is a model "productionized"?
2. How are features computed in real time to support these models?

3. How do we make sure that "what we see is what we get"? That is, minimizing the discrepancy between offline and online performance.

In most classes and in ML competitions, predictive models are pitted against a static set of hold-out data that fits in memory, and the models are allowed to take as long as they need to run. Under these lax constraints, modelers are happy to take the data, do an $O(n^3)$ operation to get features, and run it through the model. Complex feature engineering is often celebrated for pedagogical reasons. Examples:

- High-dimensionality? Don't worry, we'll just do an SVD, save off the transformation matrices, and multiply them with the hold-out data.
- Transforming arrival rate data? Hold on, let me first fit a Poisson model across the historical data and hold-out set.
- Time series? Let's throw in some Fourier coefficients.

Unfortunately, real life strips away this affordance of time and space. Predictive models are pitted against an ever-growing set of online data. In many cases, they are expected to yield predictions within milliseconds after being handed a datum. All that hard work put into building the model won't matter unless the model can handle the traffic.

Keep in mind that many models boil down to a dot product of features with weights (GLM, SVM), or a series of conjuctive statements with thresholds that can be expressed as array lookups or a series of if-else statements against the features (decision trees). So the hard part reduces to feature computation.

There are various approaches to computing features. Depending on model complexity, latency, and volume requirements, features are computed in either batch or real time. Models may choose to consume just batch features, just real-time features, or both. In some cases, a real-time model makes use of real-time features plus the output of models trained in batch.

Frameworks such as MapReduce are often used to compute features in batch. But with a more stringent latency requirement, Ian and the machine learning team at Square are working on a real-time feature computation system.

An important design goal of such a system is to ensure that historical and online features are computed in the same way. In other words, there should not be systematic differences between the online features and the historical features. Modelers should have confidence that the online performance of the models should match the expected performance.

## Data Visualization at Square

Next Ian talked about ways in which the Risk team at Square uses visualization:

- Enable efficient transaction review.
- Reveal patterns for individual customers and across customer segments.
- Measure business health.
- Provide ambient analytics.

He described a workflow tool to review users, which shows features of the seller, including the history of sales and geographical information, reviews, contact info, and more. Think mission control. This workflow tool is a type of data visualization. The operation team tasked with reviewing suspicious activity has limited time to do their work, so developers and data scientists must collaborate with the ops team to figure out ways to best represent the data. What they're trying to do with these visualizations is to augment the intelligence of the operations team, to build them an exoskeleton for identifying patterns of suspicious activity. Ian believes that this is where a lot of interesting development will emerge—the seamless combination of machine learning and data visualization.

Visualizations across customer segments are often displayed in various TVs in the office (affectionately known as "information radiators" in the Square community). These visualizations aren't necessarily trying to predict fraud per se, but rather provide a way of keeping an eye on things to look for trends and patterns over time.

This relates to the concept of "ambient analytics," which is to provide an environment for constant and passive ingestion of data so you can develop a visceral feel for it. After all, it is via the process of becoming very familiar with our data that we sometimes learn what kind of patterns are unusual or which signals deserve their own model or monitor. The Square risk team has put in a lot of effort to develop custom and generalizable dashboards.

In addition to the raw transactions, there are risk metrics that Ian keeps a close eye on. So, for example, he monitors the "clear rates" and "freeze rates" per day, as well as how many events needed to be reviewed. Using his fancy viz system he can get down to which analysts froze the most today, how long each account took to review, and what attributes indicate a long review process.

In general, people at Square are big believers in visualizing business metrics—sign-ups, activations, active users—in dashboards. They believe the transparency leads to more accountability and engagement. They run a kind of constant EKG of their business as part of ambient analytics. The risk team, in particular, are strong believers in "What gets measured gets managed."

Ian ended with his data scientist profile and a few words of advice. He thinks `plot(skill_level ~ attributes | ian)` should be shown via a log-y scale, because it doesn't take very long to be *OK* at something but it takes lots of time to get from good to great. He believes that productivity should also be measured in log-scale, and his argument is that leading software contributors crank out packages at a much higher rate than other people.

And as parting advice, Ian encourages you to:

- Play with real data.
- Build math, stats, and computer science foundations in school.
- Get an internship.
- Be literate, not just in statistics.
- Stay curious!

# Ian's Thought Experiment

Suppose you know about every single transaction in the world as it occurs. How would you use that data?

# Data Visualization for the Rest of Us

Not all of us can create data visualizations considered to be works of art worthy of museums, but it's worth building up one's ability to use data visualization to communicate and tell stories, and convey the meaning embedded in the data. Just as data science is more than a set of tools, so is data visualization, but in order to become a master, one must first be able to master the technique. Following are some tutorials and books that we have found useful in building up our data visualization skills:

- There's a nice orientation to the building blocks of data visualization by Michael Dubokov at *http://www.targetprocess.com/arti cles/visual-encoding.html*.

- Nathan Yau, who was Mark Hansen's PhD student at UCLA, has a collection of tutorials on creating visualizations in R at *http:// flowingdata.com/*. Nathan Yau also has two books: *Visualize This: The Flowing Data Guide to Design, Visualization, and Statistics* (Wiley); and *Data Points: Visualization That Means Something* (Wiley).

- Scott Murray, code artist, has a series of tutorials to get up to speed on d3 at *http://alignedleft.com/tutorials/d3/*. These have been developed into a book, *Interactive Data Visualization* (O'Reilly).

- Hadley Wickham, who developed the R package ggplot2 based on Wilkinson's Grammar of Graphics, has a corresponding book: *ggplot2: Elegant Graphics for Data Analysis (Use R!)* (Springer).

- Classic books on data visualization include several books (*The Visual Display of Quantitative Information* [Graphics Pr], for example) by Edward Tufte (a statistician widely regarded as one of the fathers of data visualization; we know we already said that about Mark Hansen—they're different generations) with an emphasis less on the tools, and more on the principles of good design. Also William Cleveland (who we mentioned back in Chapter 1 because of his proposal to expand the field of statistics into data science), has two books: *Elements of Graphing Data* (Hobart Press) and *Visualizing Data* (Hobart Press).

- Newer books by O'Reilly include the *R Graphics Cookbook*, *Beautiful Data*, and *Beautiful Visualization*.

- We'd be remiss not to mention that art schools have graphic design departments and books devoted to design principles. An education in data visualization that doesn't take these into account, as well as the principles of journalism and storytelling, and only focuses on tools and statistics is only giving you half the picture. Not to mention the psychology of human perception.

- This talk, "Describing Dynamic Visualizations" (*http://vimeo.com/66085662*) by Bret Victor comes highly recommended by Jeff Heer, a Stanford professor who created d3 with Michael Bostock (who used to work at Square, and now works for the *New York Times*). Jeff described this talk as presenting an alternative view of data visualization.

- Collaborate with an artist or graphic designer!

## Data Visualization Exercise

The students in the course, like you readers, had a wide variety of backgrounds and levels with respect to data visualization, so Rachel suggested those who felt like beginners go pick out two of Nathan Yau's tutorials and do them, and then reflect on whether it helped or not and what they wanted to do next to improve their visualization skills.

More advanced students in the class were given the option to participate in the Hubway Data Visualization challenge (*http://hubwaydata challenge.org/*). Hubway is Boston's bike-sharing program, and they released a dataset and held a competition to visualize it. The dataset is still available, so why not give it a try? Two students in Rachel's class, Eurry Kim and Kaz Sakamoto, won "best data narrative" in the competition; Rachel is very proud of them. Viewed through the lenses of a romantic relationship, their visual diary (shown in Figure 9-17) displays an inventory of their Boston residents' first 500,000 trips together.

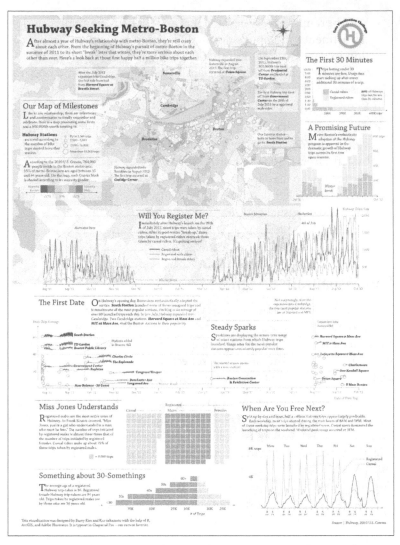

*Figure 9-17. This is a visualization by Eurry Kim and Kaz Sakamoto of the Hubway shared-bike program and its adoption by the fine people of the Boston metro area*

# CHAPTER 10
# Social Networks and Data Journalism

In this chapter we'll explore two topics that have started to become especially hot over the past 5 to 10 years: social networks and data journalism. Social networks (not necessarily just *online* ones) have been studied by sociology departments for decades, as has their counterpart in computer science, math, and statistics departments: graph theory. However, with the emergence of online social networks such as Facebook, LinkedIn, Twitter, and Google+, we now have a new rich source of data, which opens many research problems both from a social science and quantitative/technical point of view.

We'll hear first about how one company, Morningside Analytics, visualizes and finds meaning in social network data, as well as some of the underlying theory of social networks. From there, we look at constructing stories that can be told from social network data, which is a form of data journalism. Thinking of the data scientist profiles—and in this case, gene expression is an appropriate analogy—the mix of math, stats, communication, visualization, and programming required to do either data science or data journalism is slightly different, but the fundamental skills are the same. At the heart of both is the ability to ask good questions, to answer them with data, and to communicate one's findings. To that end, we'll hear briefly about data journalism from the perspective of Jon Bruner, an editor at O'Reilly.

# Social Network Analysis at Morning Analytics

The first contributor for this chapter is John Kelly from Morningside Analytics, who came to talk to us about network analysis.

Kelly has four diplomas from Columbia, starting with a BA in 1990 from Columbia College, followed by a master's, MPhil, and PhD in Columbia's School of Journalism, where he focused on network sociology and statistics in political science. He also spent a couple of terms at Stanford learning survey design and game theory and other quanty stuff. He did his master's thesis work with Marc Smith from Microsoft (*http://videolectures.net/marc_smith/*); the topic was how political discussions evolve as networks. After college and before grad school, Kelly was an artist, using computers to do sound design. He spent three years as the director of digital media at Columbia School of the Arts. He's also a programmer: Kelly taught himself Perl and Python when he spent a year in Vietnam with his wife.

Kelly sees math, statistics, and computer science (including machine learning) as tools he needs to use and be good at in order to do what he really wants to do. Like a chef in a kitchen, he needs good pots and pans and sharp knives, but the meal is the real product.

And what is he serving up in his kitchen? Kelly wants to understand how people come together, and when they do, what their impact is on politics and public policy. His company, Morningside Analytics, has clients like think tanks and political organizations. They typically want to know how social media affects and creates politics.

Communication and presentations are how he makes money—visualizations are integral to both domain expertise and communications —so his expertise lies in visualization combined with drawing conclusions from those visualizations. After all, Morningside Analytics doesn't get paid to just discover interesting stuff, but rather to help people use it.

## Case-Attribute Data versus Social Network Data

Kelly doesn't model data in the standard way through case-attribute data. Case-attribute refers to how you normally see people feed models with various "cases," which can refer to people or events—each of which have various "attributes," which can refer to age, or operating system, or search histories.

Modeling with case-attribute data started in the 1930s with early market research, and it was soon being applied to marketing as well as politics.

Kelly points out that there's been a huge bias toward modeling with case-attribute data. One explanation for this bias is that it's easy to store case-attribute data in databases, or because it's easy to *collect* this kind of data. In any case, Kelly thinks it's missing the point of the many of the questions we are trying to answer.

He mentioned Paul Lazarsfeld and Elihu Katz, two trailblazing sociologists who came here from Europe and developed the field of *social network analysis*, an approach based not only on individual people but also the relationships between them.

To get an idea of why network analysis is sometimes superior to case-attribute data analysis, think about the following example. The federal government spent money to poll people in Afghanistan. The idea was to see what citizens want in order to anticipate what's going to happen in the future. But, as Kelly points out, what'll happen isn't a simple function of what individuals think; instead, it's a question of who has the power and what *they* think.

Similarly, imagine going back in time and conducting a scientific poll of the citizenry of Europe in 1750 to determine the future politics. If you knew what you were doing you'd be looking at who was marrying whom among the royalty.

In some sense, the current focus on case-attribute data is a problem of looking for something "under the streetlamp"—a kind of observational bias wherein people are used to doing things a certain (often easier) way so they keep doing it that way, even when it doesn't answer the questions they care about.

Kelly claims that the world is a network much more than it's a bunch of cases with attributes. If you only understand how individuals behave, how do you tie things together?

# Social Network Analysis

Social network analysis comes from two places: graph theory, where Euler solved the Seven Bridges of Konigsberg problem, and sociometry, started by Jacob Moreno in the 1970s, a time when early computers were getting good at making large-scale computations on large datasets.

Social network analysis was germinated by Harrison White, professor emeritus at Columbia, contemporaneously with Columbia sociologist Robert Merton. Their idea was that people's actions have to be related to their attributes, but to really understand them you also need to look at the networks (aka systems) that *enable them to do something*.

How do we bring that idea to our models? Kelly wants us to consider what he calls the micro versus macro, or individual versus systemic divide: how do we bridge this divide? Or rather, how does this divide get bridged in various contexts?

In the US, for example, we have formal mechanisms for bridging those micro/macro divides, namely markets in the case of the "buying stuff" divide, and elections in the case of political divides. But much of the world doesn't have those formal mechanisms, although they often have a fictive shadow of those things. For the most part, we need to know enough about the actual social network to know who has the power and influence to bring about change.

# Terminology from Social Networks

The basic units of a network are called *actors* or *nodes*. They can be people, or websites, or whatever "things" you are considering, and are often indicated as a single dot in a visualization. The relationships between the actors are referred to as *relational ties* or *edges*. For example, an instance of liking someone or being friends would be indicated by an *edge*. We refer to pairs of actors as *dyads*, and triplets of actors as *triads*. For example, if we have an edge between node A and node B, and an edge between node B and node C, then *triadic closure* would be the existence of an edge between node A and node C.

We sometimes consider *subgroups*, also called *subnetworks*, which consist of a subset of the whole set of actors, along with their relational ties. Of course this means we also consider the *group* itself, which means the entirety of a "network." Note that this is a relatively easy concept in the case of, say, the Twitter network, but it's very hard in the case of "liberals."

We refer to a *relation* generally as a way of having relational ties between actors. For example, liking another person is a relation, but so is living with someone. A *social network* is the collection of some set of actors and relations.

There are actually a few different types of social networks. For example, the simplest case is that you have a bunch of actors connected by ties. This is a construct you'd use to display a Facebook graph—any two people are either friends or aren't, and any two people can theoretically be friends.

In *bipartite graphs* the connections only exist between two formally separate classes of objects. So you might have people on the one hand and companies on the other, and you might connect a person to a company if she is on the board of that company. Or you could have people and the things they're possibly interested in, and connect them if they really are.

Finally, there are *ego networks*, which is typically formed as "the part of the network surrounding a single person." For example, it could be "the subnetwork of my friends on Facebook," who may also know one another in certain cases. Studies have shown that people with higher socioeconomic status have more complicated ego networks, and you can infer someone's level of social status by looking at their ego network.

## Centrality Measures

The first question people often ask when given a social network is: *who's important here?*

Of course, there are different ways to be important, and the different definitions that attempt to capture something like importance lead to various *centrality measures*. We introduce here some of the commonly used examples.

First, there's the notion of *degree*. This counts how many people are connected to you. So in Facebook parlance, this is the number of friends you have.

Next, we have the concept of *closeness*: in words, if you are "close" to everyone, you should have a high closeness score.

To be more precise, we need the notion of distance between nodes in a *connected graph*, which in the case of a friend network means everyone is connected with everyone else through some chain of mutual friends. The distance between nodes $x$ and $y$, denoted by $d(x, y)$, is simply defined as the length of the shortest path between the two nodes. Now that you have this notation, you can define the closeness of node $x$ as the sum:

$$C(x) = \sum 2^{-d(x,y)}$$

where the sum is over all nodes $y$ distinct from $x$.

Next, there's the centrality measure called *betweenness*, which measures the extent to which people in your network know each other through you, or more precisely whether the shortest paths between them go through you. The idea here is that if you have a high betweenness score, then information probably flows through you.

To make this precise, for any two nodes $x$ and $y$ in the same connected part of a network, define $\sigma_{x,y}$ to be *the number of shortest paths between node x and node y*, and define $\sigma_{x,y}(v)$ to be *the number of shortest paths between node x and node y that go through a third node v*. Then the betweenness score of $v$ is defined as the sum:

$$B(v) = \sum \frac{\sigma_{x,y}(v)}{\sigma_{x,y}}$$

where the sum is over all distinct *pairs* of nodes $x$ and $y$ that are distinct from $v$.

The final centrality measure, which we will go into in detail in "Representations of Networks and Eigenvalue Centrality" on page 264 after we introduce the concept of an incidence matrix, is called *eigenvector centrality*. In words, a person who is *popular with the popular kids* has high eigenvector centrality. Google's PageRank is an example of such a centrality measure.

## The Industry of Centrality Measures

It's important to issue a caveat on blindly applying the preceding centrality measures. Namely, the "measurement people" form an industry in which everyone tries to sell themselves as *the* authority. But experience tells us that each has their weaknesses and strengths. The main thing is to know you're looking at the right network or subnetwork.

For example, if you're looking for a highly influential blogger in the Muslim Brotherhood, and you write down the top 100 bloggers in some large graph of bloggers, and start on the top of the list, and go down the list looking for a Muslim Brotherhood blogger, it won't work: you'll find someone who is both influential in the large network and who blogs for the Muslim Brotherhood, but they won't be influential *with* the Muslim Brotherhood, but rather with transnational elites in the larger network. In other words, you have to keep in mind the local neighborhood of the graph.

Another problem with centrality measures: experience dictates that different contexts require different tools. Something might work with blogs, but when you work with Twitter data, you'll need to get out something entirely different.

One reason is the different data, but another is the different ways people *game* centrality measures. For example, with Twitter, people create 5,000 Twitter bots that all follow one another and some strategically selected other (real) people to make them look influential by some measure (probably eigenvector centrality). But of course this isn't accurate; it's just someone gaming the measures.

Some network packages exist already and can compute the various centrality measures mentioned previously. For example, see NetworkX (*http://networkx.lanl.gov*) or igraph (*http://igraph.sourceforge.net*) if you use Python, or statnet (*http://statnet.org*) for R, or NodeXL (*http://research.microsoft.com/en-us/projects/nodexl/*), if you prefer Excel, and finally keep an eye out for a forthcoming C package from Jure Leskovec at Stanford (*http://stanford.io/18Pejdt*).

# Thought Experiment

You're part of an elite, well-funded think tank in DC. You can hire people and you have $10 million to spend. Your job is to empirically predict the future political situation of Egypt. What kinds of political parties will there be? What is the country of Egypt going to look like in 5, 10, or 20 years? You have access to exactly two of the following datasets for all Egyptians: the Facebook or Twitter network, a complete record of who went to school with who, the text or phone records of everyone, everyone's addresses, or the network data on members of all formal political organizations and private companies.

Before you decide, keep in mind that things change over time—people might migrate off of Facebook, or political discussions might need to go underground if blogging is too public. Also, Facebook alone gives a lot of information, but sometimes people will try to be stealthy—maybe the very people you are most interested in keeping tabs on. Phone records might be a better representation for that reason.

If you think this scenario is ambitious, you should know it's already being done. For example, Siemens from Germany sold Iran software to monitor their national mobile networks. In fact, governments are, generally speaking, putting more energy into loading the field with their allies, and less with shutting down the field: Pakistan hires Americans to do their pro-Pakistan blogging, and Russians help Syrians.

One last point: you should consider changing the standard direction of your thinking. A lot of the time people ask, what can we learn from this or that data source? Instead, think about it the other way around: what would it mean to predict politics in a society? And what kind of data do you need to know to do that?

In other words, figure out the questions first, and then look for the data to help answer them.

## Morningside Analytics

Kelly showed us a network map of 14 of the world's largest blogospheres. To understand the pictures, you imagine there's a force, like a wind, which sends the nodes (blogs) out to the edge, but then there's a counteracting force, namely the links between blogs, which attach them together. Figure 10-1 shows an example of the Arabic blogosphere.

The different colors represent countries and clusters of blogs. The size of each dot is centrality through degree, i.e., the number of links to other blogs in the network. The physical structure of the blogosphere can give us insight.

If we analyze text using natural language processing (NLP), thinking of the blog posts as a pile of text or a river of text, then we see the micro or macro picture only—we lose the most important story. What's missing there is social network analysis (SNA), which helps us map and analyze the patterns of interaction. The 12 different international blogospheres, for example, look different. We can infer that different societies have different interests, which give rise to different patterns.

But why are they different? After all, they're representations of some higher dimensional thing projected onto two dimensions. Couldn't it

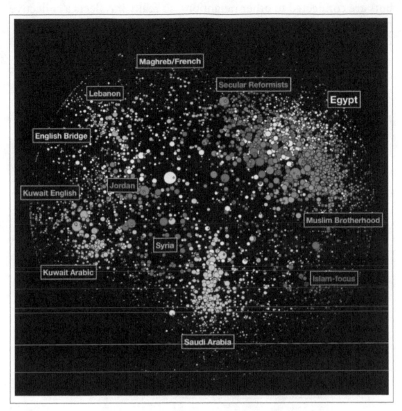

*Figure 10-1. Example of the Arabic blogosphere*

be just that they're drawn differently? Yes, but we can do lots of text analysis that convinces us these pictures really are showing us something. We put an effort into interpreting the content qualitatively.

So, for example, in the French blogosphere, we see a cluster that discusses gourmet cooking. In Germany we see various clusters discussing politics and lots of crazy hobbies. In English blogs we see two big clusters [Cathy/mathbabe interjects: gay porn and straight porn?]. They turn out to be conservative versus liberal blogs.

In Russia, their blogging networks tend to force people to stay within the networks, which is why we see very well-defined, partitioned clusters.

The proximity clustering is done using the Fruchterman-Reingold algorithm, where being in the same neighborhood means your neighbors are connected to other neighbors, so really it reflects a collective phenomenon of influence. Then we interpret the segments. Figure 10-2 shows an example of English language blogs.

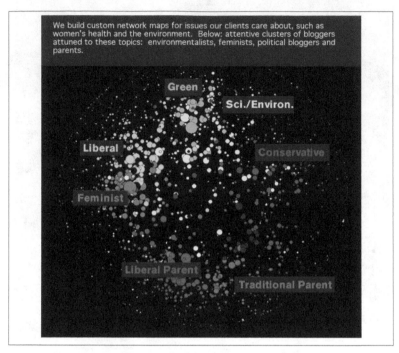

*Figure 10-2. English language blogs*

## How Visualizations Help Us Find Schools of Fish

Social media companies are each built around the fact that they either have the data or they have a toolkit—a patented sentiment engine or something, a *machine that goes ping*. Keep in mind, though, that social media is heavily a product of organizations that pay to move the needle —that is, that *game the machine that goes ping*. To believe what you see, you need to keep ahead of the game, which means you need to decipher that game to see how it works. That means you need to visualize.

Example: if you are thinking about elections, look at people's blogs within "moms" or "sports fans." This is more informative than looking at partisan blogs where you already know the answer.

Here's another example: Kelly walked us through an analysis, after binning the blogosphere into its segments, of various types of links to partisan videos like MLK's "I Have a Dream" speech, and a video from the Romney campaign. In the case of the MLK speech, you see that it gets posted in spurts around the election cycle events all over the blogosphere, but in the case of the Romney campaign video, you see a concerted effort by conservative bloggers to post the video in unison.

That is to say, if you were just looking at a histogram of links—a pure count—it might look as if the Romney video had gone viral, but if you look at it through the lens of the understood segmentation of the blogosphere, it's clearly a planned operation to game the "virality" measures.

Kelly also works with the Berkman Center for Internet and Society at Harvard. He analyzed the Iranian blogosphere in 2008 and again in 2011, and he found much the same in terms of clustering—young anti-government democrats, poetry (an important part of Iranian culture), and conservative pro-regime clusters dominated in both years.

However, only 15% of the blogs are the same from 2008 to 2011.

So, whereas people are often concerned about *individuals* (the case-attribute model), the individual fish are less important than the *schools* of fish. By doing social network analysis, we are looking for the schools, because that way we learn about the salient interests of the society and how those interests are stable over time.

The moral of this story is that we need to focus on meso-level patterns, not micro- or macro-level patterns.

# More Background on Social Network Analysis from a Statistical Point of View

One way to start with SNA is to think about a network itself as a random object, much like a random number or random variable. The network can be conceived of as the result of a random process or as coming from an underlying probability distribution. You can in fact imagine a sample of networks, in which case you can ask questions like: What characterizes networks that might conceivably be Twitter-like? Could a given network reflect real-world friendships? What would it even mean to say yes or no to this question?

These are some of the basic questions in the discipline of social network analysis, which has emerged from academic fields such as math, statistics, computer science, physics, and sociology, with far-ranging applications in even more fields including fMRI research, epidemiology, and studies of online social networks such as Facebook or Google+.

## Representations of Networks and Eigenvalue Centrality

In some networks, the edges between nodes are directed: I can follow you on Twitter when you don't follow me, so there will be an edge *from* me *to* you. But other networks have only symmetric edges: we either know each other or don't. These latter types of networks are called *undirected*.

An undirected network with $N$ nodes can be represented by an $N \times N$ matrix comprised of 1s and 0s, where the $(i,j)$th element in the matrix is a 1 if and only if nodes $i$ and $j$ are connected. This matrix is known as an *adjacency matrix*, or *incidence matrix*. Note that we can actually define this for directed networks too, but for undirected networks, the matrix is always symmetric.

Alternatively, a network can be represented by a list of lists: for each node $i$, we list the nodes to which node $i$ is connected. This is known as an *incidence* list, and note that it doesn't depend on the network being undirected. Representing the network this way saves storage space—the nodes can have attributes represented as a vector or list. For example, if the nodes are people, the attributes could be demographic information or information about their behavior, habits, or tastes.

The edges themselves can also have values, or weights/vectors, which could capture information about the nature of the relationship between the nodes they connect. These values could be stored in the $N \times N$ matrix, in place of the 1s and 0s that simply represent the presence or not of a relationship.

Now with the idea of an adjacency matrix $A$ in mind, we can finally define *eigenvalue centrality*, which we first mentioned in "Centrality Measures" on page 257. It is compactly defined as the unique vector solution $x$ to the equation:

$$Ax = \lambda x$$

such that

$$x_i > 0, \quad i = 1 \cdots N$$

As it turns out, that last condition is equivalent to choosing the largest eigenvalue $\lambda$. So for an actual algorithm, find the roots of the equation $det(A - tI)$ and order them by size, grabbing the biggest one and calling it $\lambda$. Then solve for $x$ by solving the system of equations:

$$(A - \lambda I)x = 0$$

Now we have $x$, the vector of eigenvector centrality scores.

Note this doesn't give us much of a *feel* for eigenvalue centrality, even if it gives us a way to compute it. You can get that feel by thinking about it as the limit of a simple iterative scheme—although it requires proof, which you can find, for example, here (*http://goo.gl/UVkLoF*).

Namely, start with a vector whose entries are just the degrees of the nodes, perhaps scaled so that the sum of the entries is 1. The degrees themselves aren't giving us a real understanding of how interconnected a given node is, though, so in the next iteration, add the degrees of all the neighbors of a given node, again scaled. Keep iterating on this, adding degrees of neighbors one further step out each time. In the limit as this iterative process goes on forever, we'll get the eigenvalue centrality vector.

## A First Example of Random Graphs: The Erdos-Renyi Model

Let's work out a simple example where a network can be viewed as a single realization of an underlying stochastic process. Namely, where the existence of a given edge follows a probability distribution, and *all the edges are considered independently*.

Say we start with $N$ nodes. Then there are $D = \binom{N}{2}$ pairs of nodes, or *dyads*, which can either be connected by an (undirected) edge or not. Then there are $2^D$ possible observed networks. The simplest underlying distribution one can place on the individual edges is called the *Erdos-Renyi model*, which assumes that for every pair of nodes $(i, j)$, an edge exists between the two nodes with probability $p$.

## The Bernoulli Network

Not all networks with $N$ nodes occur with equal probability under this model: observing a network with all nodes attached to all other nodes has probability $p^D$, while observing a network with all nodes disconnected has probability $(1-p)^D$. And of course there are many other possible networks between these two extremes. The Erdos-Renyi model is also known as a *Bernoulli network*. In the mathematics literature, the Erdos-Renyi model is treated as a mathematical object with interesting properties that allow for theorems to be proved.

# A Second Example of Random Graphs: The Exponential Random Graph Model

Here's the bad news: social networks that can be observed in the real world tend not to resemble Bernoulli networks. For example, friendship networks or academic collaboration networks commonly exhibit characteristics such as *transitivity* (the tendency, when A knows B and B knows C, that A knows C), clustering (the tendency for more or less well-defined smallish groups to exist with larger networks), reciprocity or mutuality (in a directed network, the tendency for A to follow B if B follows A), and betweenness (the tendency for there to exist special people through whom information flows).

Some of these observed properties of real-world networks are pretty simple to translate into mathematical language. For example, transitivity can be captured by the number of triangles in a network.

Exponential random graph models (ERGMs) are an approach to capture these real-world properties of networks, and they are commonly used within sociology.

The general approach for ERGMs is to choose pertinent *graph statistics* like the number of triangles, the number of edges, and the number of *2-stars* (subgraphs consisting of a node with two spokes—so a node with degree 3 has three *2-stars* associated to it) given the number of nodes, and have these act as variables $z_i$ of your model, and then tweak the associated coefficients $\theta_i$ to get them tuned to a certain type of behavior you observe or wish to simulate. If $z_1$ refers to the number of triangles, then a positive value for $\theta_1$ would indicate a tendency toward a larger number of triangles, for example.

Additional graph statistics that have been introduced include $k$-stars (subgraphs consisting of a node with $k$ spokes—so a node with degree $k+1$ has $k+1$ $k$-stars associated with it), degree, or *alternating $k$-stars*, an aggregation statistics on the number of $k$-stars for various $k$. Let's give you an idea of what an ERGM might look like formula-wise:

$$Pr(Y = y) = \left(\frac{1}{\kappa}\right)\left((\theta_1 z_1(y) + \theta_2 z_2(y) + \theta_3 z_3(y)\right)$$

Here we're saying that the probability of observing one particular realization of a random graph or network, $Y$, is a function of the graph statistics or properties, which we just described as denoted by $z_i$.

In this framework, a Bernoulli network is a special case of an ERGM, where we only have one variable corresponding to number of edges.

### Inference for ERGMs

Ideally—though in some cases unrealistic in practice—one could observe a sample of several networks, $Y_1, ..., Y_n$, each represented by their adjacency matrices, say for a fixed number $N$ of nodes.

Given those networks, we could model them as independent and identically distributed observations from the same probability model. We could then make inferences about the parameters of that model.

As a first example, if we fix a Bernoulli network, which is specified by the probability $p$ of the existence of any given edge, we can calculate the likelihood of any of our sample networks having come from that Bernoulli network as

$$L = \Pi_i^n p^{d_i}(1-p)^{D-d_i}$$

where $d_i$ is the number of observed edges in the $i$th network and $D$ is the total number of dyads in the network, as earlier. Then we can back out an estimator for $p$ as follows:

$$\hat{p} = \frac{\sum_{i=1}^n d_i}{nD}$$

In practice in the ERGM literature, only one network is observed, which is to say *we work with a sample size of one*. From this one example

---

we estimate a parameter for the probability model that "generated" this network. For a Bernoulli network, from even just one network, we could estimate $p$ as the proportion of edges out of the total number of dyads, which seems a reasonable estimate.

But for more complicated ERGMs, estimating the parameters from one observation of the network is tough. If it's done using something called a *pseudo-likelihood estimation procedure*, it sometimes produces infinite values (see Mark Handcock's 2003 paper, "Assessing Degeneracy of Statistical Models of Social Networks"). If it's instead done using something called *MCMC methods*, it suffers from something called *inferential degeneracy*, where the algorithms converge to degenerate graphs—graphs that are complete or empty—or the algorithm does not converge consistently (also covered in Handcock's paper).

### Further examples of random graphs: latent space models, small-world networks

Motivated by problems of model degeneracy and instability in exponential random graph models, researchers introduced *latent space models* (see Peter Hoff's "Latent Space Approaches to Social Network Analysis").

Latent space models attempt to address the following issue: we observe some reality, but there is some corresponding latent reality that we cannot observe. So, for example, we may observe connections between people on Facebook, but we don't observe where those people live, or other attributes that make them have a tendency to befriend each other.

Other researchers have proposed *small-world networks* (see the Watts and Strogatz model proposed in their 1998 paper), which lie on a spectrum between completely random and completely regular graphs and attempt to capture the real-world phenomenon of six degrees of separation. A criticism of this model is that it produces networks that are homogeneous in degree, whereas observable real-world networks tend to be scale-free and inhomogeneous in degree.

In addition to the models just described, other classes of models include Markov random fields, stochastic block models, mixed membership models, and stochastic block mixed membership models— each of which model relational data in various ways, and seek to include properties that other models do not. (See, for example, the paper "Mixed Membership Stochastic Block Models" by Edoardo Airoli, et al.)

Here are some textbooks for further reading:

- *Networks, Crowds, and Markets* (Cambridge University Press) by David Easley and Jon Kleinberg at Cornell's computer science department.

- Chapter on Mining Social-Network graphs in the book *Mining Massive Datasets* (Cambridge University Press) by Anand Rajaraman, Jeff Ullman, and Jure Leskovec in Stanford's computer science department.

- *Statistical Analysis of Network Data* (Springer) by Eric D. Kolaczyk at Boston University.

# Data Journalism

Our second speaker of the night was Jon Bruner, an editor at O'Reilly who previously worked as the data editor at Forbes. He is broad in his skills: he does research and writing on anything that involves data.

## A Bit of History on Data Journalism

Data journalism has been around for a while, but until recently, computer-assisted reporting was a domain of Excel power users. (Even now, if you know how to write an Excel program, you're an elite.)

Things started to change recently: more data became available to us in the form of APIs, new tools, and less expensive computing power —so almost anyone can analyze pretty large datasets on a laptop. Programming skills are now widely enough held so that you can find people who are both good writers and good programmers. Many people who are English majors know enough about computers to get by; or on the flip side, you'll find computer science majors who can write.

In big publications like the *New York Times*, the practice of data journalism is divided into fields: graphics versus interactive features, research, database engineers, crawlers, software developers, and domain-expert writers. Some people are in charge of raising the right questions but hand off to others to do the analysis. Charles Duhigg at the *New York Times*, for example, studied water quality in New York, and got a Freedom of Information Act request to the State of New York

—he knew enough to know what would be in that FOIA request and what questions to ask, but someone else did the actual analysis.

At a smaller organization, things are totally different. Whereas the *New York Times* has 1,000 people on its newsroom "floor," *The Economist* has maybe 130, and *Forbes* has 70 or 80 people in its newsrooms. If you work for anything besides a national daily, you end up doing everything by yourself: you come up with a question, you go get the data, you do the analysis, then you write it up. (Of course, you can also help and collaborate with your colleagues when possible.)

## Writing Technical Journalism: Advice from an Expert

Jon was a math major in college at the University of Chicago, after which he took a job writing at *Forbes*, where he slowly merged back into quantitative work. For example, he found himself using graph theoretic tools when covering contributions of billionaires to politicians.

He explained the term "data journalism" to the class by way of explaining his own data scientist profile.

First of all, it involved *lots* of data visualization, because it's a fast way of describing the bottom line of a dataset. Computer science skills are pretty important in data journalism, too. There are tight deadlines, and the data journalist has to be good with their tools and with messy data—because even federal data is messy. One has to be able to handle arcane formats, and often this means parsing stuff in Python. Jon himself uses JavaScript, Python, SQL, and MongoDB, among other tools.

Statistics, Bruno says, informs the way you think about the world. It inspires you to write things: e.g., the *average person* on Twitter is a woman with 250 followers, but the *median person* has 0 followers—the data is clearly skewed. That's an inspiration right there for a story.

Bruno admits to being a novice in the field of machine learning. However, he claims domain expertise as critical in data journalism: with exceptions for people who can specialize in one subject, say at a governmental office or a huge daily, for a smaller newspaper you need to be broad, and you need to acquire a baseline layer of expertise quickly.

Of course communications and presentations are absolutely huge for data journalists. Their fundamental skill is *translation*: taking complicated stories and deriving meaning that readers will understand. They also need to anticipate questions, turn them into quantitative experiments, and answer them persuasively.

Here's advice from Jon for anyone initiating a data journalism project: *don't have a strong thesis before you interview the experts.* Go in with a loose idea of what you're searching for and be willing to change your mind and pivot if the experts lead you in a new and interesting direction. Sounds kind of like exploratory data analysis!

# CHAPTER 11

# Causality

Many of the models and examples in the book so far have been focused on the fundamental problem of prediction. We've discussed examples like in Chapter 8, where your goal was to build a model to predict whether or not a person would be likely to prefer a certain item—a movie or a book, for example. There may be thousands of features that go into the model, and you may use feature selection to narrow those down, but ultimately the model is getting optimized in order to get the highest accuracy. When one is optimizing for accuracy, one doesn't necessarily worry about the *meaning* or *interpretation* of the features, and especially if there are thousands of features, it's well-near impossible to interpret at all.

Additionally, you wouldn't even want to make the statement that certain characteristics *caused* the person to buy the item. So, for example, your model for predicting or recommending a book on Amazon could include a feature "whether or not you've read Wes McKinney's O'Reilly book *Python for Data Analysis*." We wouldn't say that reading his book *caused* you to read *this* book. It just might be a good predictor, which would have been discovered and come out as such in the process of optimizing for accuracy. We wish to emphasize here that it's not simply the familiar correlation-causation trade-off you've perhaps had drilled into your head already, but rather that your *intent* when building such a model or system was not even to understand causality at all, but rather to *predict*. And that if your intent *were* to build a model that helps you get at causality, you would go about that in a different way.

A whole different set of real-world problems that actually use the same statistical methods (logistic regression, linear regression) as part of the building blocks of the solution are situations where you *do* want to understand causality, when you want to be able to say that a certain type of behavior *causes* a certain outcome. In these cases your mentality or goal is not to optimize for predictive accuracy, but rather to be able to isolate causes.

This chapter will explore the topic of causality, and we have two experts in this area as guest contributors, Ori Stitelman and David Madigan. Madigan's bio will be in the next chapter and requires this chapter as background. We'll start instead with Ori, who is currently a data scientist at Wells Fargo. He got his PhD in biostatistics from UC Berkeley after working at a litigation consulting firm. As part of his job, he needed to create stories from data for experts to testify at trial, and he thus developed what he calls "data intuition" from being exposed to tons of different datasets.

# Correlation Doesn't Imply Causation

One of the biggest statistical challenges, from both a theoretical and practical perspective, is establishing a causal relationship between two variables. When does one thing cause another? It's even trickier than it sounds.

Let's say we discover a correlation between ice cream sales and bathing suit sales, which we display by plotting ice cream sales and bathing suit sales over time in Figure 11-1.

This demonstrates a close association between these two variables, but it doesn't establish *causality*. Let's look at this by pretending to know nothing about the situation. All sorts of explanations might work here. Do people find themselves irrestistably drawn toward eating ice cream when they wear bathing suits? Do people change into bathing suits every time they eat ice cream? Or is there some third thing (like hot weather) which we haven't considered that causes both? Causal inference is the field that deals with better understanding the conditions under which association can be interpreted as causality.

## Asking Causal Questions

The natural form of a causal question is: What is the effect of $x$ on $y$?

Some examples are: "What is the effect of *advertising* on *customer behavior*?" or "What is the effect of *drug* on *time until viral failure*?" or in the more general case, "What is the effect of *treatment* on *outcome*?"

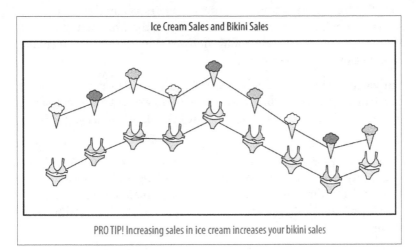

Figure 11-1. Relationship between ice cream sales and bathing suit sales

 The terms "treated" and "untreated" come from the biostatistics, medical, and clinical trials realm, where patients are given a medical treatment, examples of which we will encounter in the next chapter. The terminology has been adopted by the statistical and social science literature.

It turns out estimating causal parameters is hard. In fact, the effectiveness of advertising is almost always considered a moot point because it's so hard to measure. People will typically choose metrics of success that are easy to estimate but don't measure what they want, and everyone makes decisions based on them anyway because it's easier. But they have real negative effects. For example, marketers end up being rewarded for selling stuff to people online who would have bought something anyway.

## Confounders: A Dating Example

Let's look at an example from the world of online dating involving a lonely guy named Frank. Say Frank is perusing a dating website and comes upon a very desirable woman. He wants to convince her to go out with him on a date, but first he needs to write an email that will get her interested. What should he write in his email to her? Should he tell her she is beautiful? How do we test that with data?

Let's think about a randomized experiment Frank could run. He could select a bunch of beautiful women, and half the time, randomly, tell them they're beautiful. He could then see the difference in response rates between the two groups.

For whatever reason, though, Frank doesn't do this—perhaps he's too much of a romantic—which leaves us to try to work out whether saying a woman is beautiful is a good move for Frank. It's on us to get Frank a date.

If we could, we'd understand the future under two alternative realities: the reality where he sends out the email telling a given woman she's beautiful and the reality where he sends an email but doesn't use the word beautiful. But only one reality is possible. So how can we proceed?

Let's write down our causal question explicitly: what is the effect of Frank telling a woman she's beautiful on him getting a positive response?

In other words, the "treatment" is Frank's telling a woman she's beautiful over email, and the "outcome" is a positive response in an email, or possibly no email at all. An email from Frank that doesn't call the recipient of the email beautiful would be the control for this study.

There are lots of things we're not doing here that we might want to try. For example, we're not thinking about Frank's attributes. Maybe he's a really weird unattractive guy that no woman would want to date no matter what he says, which would make this a tough question to solve. Maybe he can't even spell "beautiful." Conversely, what if he's gorgeous and/or famous and it doesn't matter what he says? Also, most dating sites allow women to contact men just as easily as men contact women, so it's not clear that our definitions of "treated" and "untreated" are well-defined. Some women might ignore their emails but spontaneously email Frank anyway.

## OK Cupid's Attempt

As a first pass at understanding the impact of word choice on response rates, the online dating site OK Cupid analyzed over 500,000 first contacts on its site. They looked at keywords and phrases, and how they affected reply rates, shown in Figure 11-2.

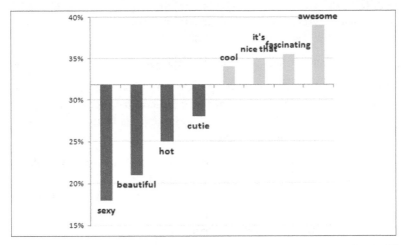

*Figure 11-2. OK Cupid's attempt to demonstrate that using the word "beautiful" in an email hurts your chances of getting a response*

The y-axis shows the response rate. On average the response rate across *all* emails was ~32%. They then took the subset of emails that included a certain word such as "beautiful" or "awesome," and looked at the response rate for those emails. Writing this in terms of conditional probabilities, we would say they were estimating these: P(response) = 0.32 vs P(response|"beautiful") = 0.22.

One important piece of information missing in this plot is the bucket sizes. How many first contacts contained each of the words? It doesn't really change things, except it would help in making it clear that the horizontal line at 32% is a weighted average across these various buckets of emails.

They interpreted this graph and created a rule called "Avoid Physical Compliments." They discussed this in the blog post "Exactly what to say in a first message" with the following explanation: "You might think that words like gorgeous, beautiful, and sexy are nice things to say to someone, but no one wants to hear them. As we all know, people normally like compliments, but when they're used as pick-up lines, before you've even met in person, they inevitably feel… ew. Besides, when you tell a woman she's beautiful, chances are you're not."

This isn't an experiment, but rather an *observational study*, which we'll discuss more later but for now means we collect data as it naturally occurs in the wild. Is it reasonable to conclude from looking at this plot that adding "awesome" to an email increases the response rate, or that "beautiful" decreases the response rate?

Before you answer that, consider the following three things.

First, it could say more about the *person* who says "beautiful" than the word itself. Maybe they are otherwise ridiculous and overly sappy? Second, people may be describing *themselves* as beautiful, or some third thing like the world we live in.

These are both important issues when we try to understand population-wide data such as in the figure, because they address the question of whether having the word "beautiful" in the body of the email actually implies what we think it does. But note that both of those issues, if present, are consistently true for a given dude like Frank trying to get a date. So if Frank is sappy, he's theoretically equally sappy to all the women he writes to, which makes it a consistent experiment, from his perspective, to decide whether or not to use the word "beautiful" in his emails.

The third and most important issue to consider, because it does *not* stay consistent for a given dude, is that *the specific recipients of emails containing the word "beautiful" might be special*: for example, they might get tons of email, and only respond to a few of them, which would make it less likely for Frank to get any response at all.

In fact, if the woman in question is beautiful (let's pretend that's a well-defined term), that fact affects two separate things at the same time. Both whether Frank uses the word "beautiful" or not in his email, and the outcome, i.e., whether Frank gets a response. For this reason, the fact that the woman is beautiful qualifies as a *confounder*; in other words, a variable that influences or has a causal effect on both the treatment itself as well as the outcome.

Let's be honest about what this plot *actually* shows versus what OK Cupid was implying it showed. It shows the observed response rate for emails that contained the given words. It should *not* be used and cannot correctly be interpreted as a prescription or suggestion for how to construct an email to get a response because after *adjusting* for confounders, which we'll discuss later in the chapter, using the word "beautiful" could be the best thing we could do. We can't say for sure because we don't have the data, but we'll describe what data we'd need and how we'd analyze it to do this study properly. Their advice might be correct, but the plot they showed does not back up this advice.

# The Gold Standard: Randomized Clinical Trials

So what do we do? How do people *ever* determine causality?

The gold standard for establishing causality is the randomized experiment. This a setup whereby we randomly assign some group of people to receive a "treatment" and others to be in the "control" group —that is, they *don't* receive the treatment. We then have some outcome that we want to measure, and the causal effect is simply the difference between the treatment and control group in that measurable outcome. The notion of using experiments to estimate causal effects rests on the statistical assumption that using randomization to select two groups has created "identical" populations from a statistical point of view.

Randomization works really well: because we're flipping coins, all other factors that might be confounders (current or former smoker, say) are more or less removed, because we can guarantee that smokers will be fairly evenly distributed between the two groups if there are enough people in the study.

The truly brilliant thing about randomization is that randomization matches well on the possible confounders we thought of, but will also give us balance on the *50 million things we didn't think of.*

So, although we can algorithmically find a better split for the ones we thought of, that quite possibly wouldn't do as well on the other things. That's why we really do it randomly, because it does quite well on things we think of and things we don't.

But there's bad news for randomized clinical trials as well, as we pointed out earlier. First off, it's only ethically feasible if there's something called clinical equipoise, which means the medical community really doesn't know which treatment is better. If we know treating someone with a drug will be better for them than giving them nothing, we can't randomly not give people the drug.

For example, if we want to tease out the relationship between smoking and heart disease, we can't randomly assign someone to smoke, because it's known to be dangerous. Similarly, the relationship between cocaine and birthweight is fraught with danger, as is the tricky relationship between diet and mortality.

The other problem is that they are expensive and cumbersome. It takes a long time and lots of people to make a randomized clinical trial work. On the other hand, not doing randomized clinical trials can lead to mistaken assumptions that are extremely expensive as well.

Sometimes randomized studies are just plain unfeasible. Let's go back to our OK Cupid example, where we have a set of observational data and we have a good reason to believe there are confounders that are screwing up our understanding of the effect size. As noted, the gold standard would be to run an experiment, and while the OK Cupid employees *could* potentially run an experiment, it would be unwise for them to do so—randomly sending email to people telling them they are "beautiful" would violate their agreement with their customers.

In conclusion, *when they are possible*, randomized clinical trials are the gold standard for elucidating cause-and-effect relationships. It's just that they aren't always possible.

---

## Average Versus the Individual

Randomized clinical trials measure the effect of a certain drug averaged across all people. Sometimes they might bucket users to figure out the average effect on men or women or people of a certain age, and so on. But in the end, it still has averaged out stuff so that for a given individual we don't know what they effect would be on them. There is a push these days toward personalized medicine with the availability of genetic data, which means we stop looking at averages because we want to make inferences about the one. Even when we were talking about Frank and OK Cupid, there's a difference between conducting this study across all men versus Frank alone.

---

# A/B Tests

In software companies, what we described as random experiments are sometimes referred to as A/B tests. In fact, we found that if we said the word "experiments" to software engineers, it implied to them "trying something new" and not necessarily the underlying statistical design of having users experience different versions of the product in order to measure the impact of that difference using metrics. The concept is intuitive enough and seems simple. In fact, if we set up the infrastructure properly, running an experiment can come down to writing a short configuration file and changing just one parameter—be it a different color or layout or underlying algorithm—that gives some users a different experience than others. So, there are aspects of running A/B tests in a tech company that make it much easier than in a clinical trial. And there's much less at stake in terms of the idea that we're not dealing with people's lives. Other convenient things are there aren't compliance issues, so with random clinical trials we can't control whether someone takes the drug or not, whereas online, we can control what we show the user. But notice we said *if* we set up the experimental infrastructure properly, and that's a big IF.

It takes a lot of work to set it up well and then to properly analyze the data. When different teams at a company are all working on new features of a product and all want to try out variations, then if you're not careful a single user could end up experiencing multiple changes at once. For example, the UX team might change the color or size of the font, or the layout to see if that increases click-through rate. While at the same time the content ranking team might want to change the algorithm that chooses what to recommend to users, and the ads team might be making changes to their bidding system. Suppose the metric you care about is return rate, and a user starts coming back more, and you had them in three different treatments but you didn't know that because the teams weren't coordinating with each other. Your team might assume the treatment is the reason the user is coming back more, but it might be the combination of all three.

There are various aspects of an experimental infrastructure that you need to consider, which are described in much more detail in Overlapping Experiment Infrastructure: More, Better, Faster Experimentation, a 2010 paper by Google employees Diane Tang, et al. See the following sidebar for an excerpt from this paper.

# From "Overlapping Experiment Infrastructure: More, Better, Faster Experimentation"

The design goals for our experiment infrastructure are therefore: more, better, faster.

*More*

> We need scalability to run more experiments simultaneously. However, we also need flexibility: different experiments need different configurations and different sizes to be able to measure statistically significant effects. Some experiments only need to change a subset of traffic, say Japanese traffic only, and need to be sized appropriately. Other experiments may change all traffic and produce a large change in metrics, and so can be run on less traffic.

*Better*

> Invalid experiments should not be allowed run on live traffic. Valid but bad experiments (e.g., buggy or unintentionally producing really poor results) should be caught quickly and disabled. Standardized metrics should be easily available for all experiments so that experiment comparisons are fair: two experimenters should use the same filters to remove robot traffic when calculating a metric such as CTR.

*Faster*

> It should be easy and quick to set up an experiment; easy enough that a non-engineer can do so without writing any code. Metrics should be available quickly so that experiments can be evaluated quickly. Simple iterations should be quick to do. Ideally, the system should not just support experiments, but also controlled ramp-ups, i.e., gradually ramping up a change to all traffic in a systematic and well-understood way.

That experimental infrastructure has a large team working on it and analyzing the results of the experiments on a full-time basis, so this is nontrivial. To make matters more complicated, now that we're in an age of social networks, we can no longer assume that users are independent (which is part of the randomization assumption underlying experiments). So, for example, Rachel might be in the treatment group of an experiment Facebook is running (which is impossible because Rachel isn't actually on Facebook, but just pretend), which lets Rachel post some special magic kind of post, and Cathy might be in the control group, but she still sees the special magic post, so she actually received a different version of the treatment, so the experimental design must take into account the underlying network structure. This is a nontrivial problem and still an open research area.

# Second Best: Observational Studies

While the gold standard is generally understood to be randomized experiments or A/B testing, they might not always be possible, so we sometimes go with second best, namely observational studies.

Let's start with a definition:

> An observational study is an empirical study in which the objective is to elucidate cause-and-effect relationships in which it is not feasible to use controlled experimentation.

Most data science activity revolves around observational data, although A/B tests, as you saw earlier, are exceptions to that rule. Most of the time, the data you have is what you get. You don't get to replay a day on the market where Romney won the presidency, for example.

Designed studies are almost always theoretically better tests, as we know, but there are plenty of examples where it's unethical to run them. Observational studies are done in contexts in which you can't do designed experiments, in order to elucidate cause-and-effect.

In reality, sometimes you don't care about cause-and-effect; you just want to build predictive models. Even so, there are many core issues in common with the two.

## Simpson's Paradox

There are all kinds of pitfalls with observational studies.

For example, look at the graph in Figure 11-3, where you're finding a best-fit line to describe whether taking higher doses of the "bad drug" is correlated to higher probability of a heart attack.

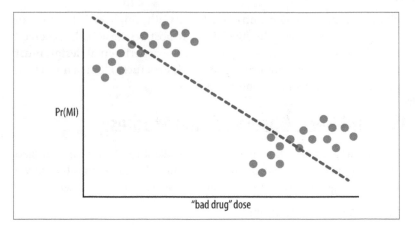

*Figure 11-3. Probability of having a heart attack (also known as MI, or myocardial infarction) as a function of the size of the dose of a bad drug*

It looks like, from this vantage point, the higher the dose, the fewer heart attacks the patient has. But there are two clusters, and if you know more about those two clusters, you find the opposite conclusion, as you can see in Figure 11-4.

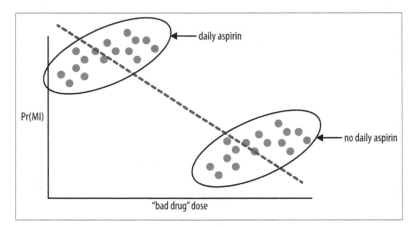

*Figure 11-4. Probability of having a heart attack as a function of the size of the dose of a bad drug and whether or not the patient also took aspirin*

This picture was rigged, so the issue is obvious. But, of course, when the data is multidimensional, you wouldn't even always draw such a simple picture.

In this example, we'd say aspirin-taking is a confounder. We'll talk more about this in a bit, but for now we're saying that the aspirin-taking or nonaspirin-taking of the people in the study wasn't randomly distributed among the people, and it made a huge difference in the apparent effect of the drug.

Note that, if you think of the original line as a predictive model, it's actually *still* the best model you can obtain knowing nothing more about the aspirin-taking habits or genders of the patients involved. The issue here is really that you're trying to assign causality.

It's a general problem with regression models on observational data. You have no idea what's going on. As Madigan described it, "it's the Wild West out there."

It could be the case that within each group there are males and females, and if you partition by *those*, you see that the more drugs they take, the better again. Because a given person either is male or female, and either takes aspirin or doesn't, this kind of thing really matters.

This illustrates the fundamental problem in observational studies: a trend that appears in different groups of data disappears when these groups are combined, or vice versa. This is sometimes called Simpson's Paradox.

## The Rubin Causal Model

The Rubin causal model is a mathematical framework for understanding what information we know and don't know in observational studies.

It's meant to investigate the confusion when someone says something like, "I got lung cancer because I smoked." Is that true? If so, you'd have to be able to support the statement, "If I hadn't smoked, I wouldn't have gotten lung cancer," but nobody knows that for sure.

Define $Z_i$ to be the treatment applied to unit $i$ (0 = control, 1 = treatment), $Y_i(1)$ to be the response for unit $i$ if $Z_i = 1$ and $Y_i(0)$ to be the response for unit $i$ if $Z_i = 0$.

Then the *unit level causal effect*, the thing we care about, is $Y_i(1) - Y_i(0)$, but we only see one of $Y_i(0)$ and $Y_i(1)$.

Example: $Z_i$ is 1 if I smoked, 0 if I didn't (I am the unit). $Y_i(1)$ is 1 if I got cancer and I smoked, and 0 if I smoked and didn't get cancer. Similarly $Y_i(0)$ is 1 or 0, depending on whether I got cancer while not smoking. The overall causal effect on me is the difference $Y_i(1) - Y_i(0)$. This is equal to 1 if I really got cancer because I smoked, it's 0 if I got cancer (or didn't) independent of smoking, and it's $-1$ if I avoided cancer by smoking. But I'll never know my actual value because I only know one term out of the two.

On a population level, we do know how to infer that there are quite a few "1"s among the population, but *we will never be able to assign a given individual that number.*

This is sometimes called the fundamental problem of causal inference.

## Visualizing Causality

We can represent the concepts of causal modeling using what is called a *causal graph.*

Denote by $W$ the set of all potential confounders. Note it's a big assumption that we can take account of all of them, and we will soon see how unreasonable this seems to be in epidemiology research in the next chapter.

In our example with Frank, we have singled out one thing as a potential confounder—the woman he's interested in being beautiful—but if we thought about it more we might come up with other confounders, such as whether Frank is himself attractive, or whether he's desperate, both of which affect how he writes to women as well as whether they respond positively to him.

Denote by $A$ the treatment. In our case the treatment is Frank's using the word "beautiful" in an introductory email. We usually assume this to have a binary (0/1) status, so for a given woman Frank writes to, we'd assign her a "1" if Frank uses the word "beautiful." Just keep in mind that if he says it's beautiful weather, we'd be measuring counting that as a "1" even though we're thinking about him calling the woman beautiful.

Denote by $Y$ the binary (0/1) outcome. We'd have to make this well-defined, so, for example, we can make sure Frank asks the women he writes to for their phone number, and we could define a positive outcome, denoted by "1," as Frank getting the number. We'd need to make this as precise as possible, so, for example, we'd say it has to happen in the OK Cupid platform within a week of Frank's original email. Note we'd be giving a "1" to women who ignore his emails but for some separate reason send him an email with their number. It would also be hard to check that the number isn't fake.

The nodes in a causal graph are labeled by these sets of confounders, treatment, and outcome, and the directed edges, or arrows, indicate causality. In other words, the node the arrow is coming out of in some way directly affects the node the arrow is going into.

In our case we have Figure 11-5.

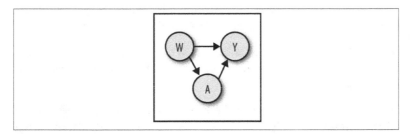

*Figure 11-5. Causal graph with one treatment, one confounder, and one outcome*

In the case of the OK Cupid example, the causal graph is the simplest possible causal graph: one treatment, one confounder, and one outcome. But they can get much more complicated.

## Definition: The Causal Effect

Let's say we have a population of 100 people that take some drug, and we screen them for cancer. Say 30 of them get cancer, which gives them a cancer rate of 0.30. We want to ask the question, did the drug cause the cancer?

To answer that, we'd have to know what would've happened if they hadn't taken the drug. Let's play God and stipulate that, had they not taken the drug, we would have seen 20 get cancer, so a rate of 0.20. We typically measure the increased risk of cancer as the difference of these two numbers, and we call it the *causal effect*. So in this case, we'd say the causal effect is 10 percentage points.

The *causal effect* is sometimes defined as the *ratio* of these two numbers instead of the difference.

But we don't have God's knowledge, so instead we choose another population to compare this one to, and we see whether *they* get cancer or not, while *not* taking the drug. Say they have a natural cancer rate of 0.10. Then we would conclude, using them as a proxy, that the increased cancer rate is the difference between 0.30 and 0.10, so 20%. This is of course wrong, but the problem is that the two populations have some underlying differences that we don't account for.

If these were the "same people," down to the chemical makeup of each others' molecules, this proxy calculation would work perfectly. But of course they're not.

So how do we actually select these people? One technique is to use what is called propensity score matching or modeling. Essentially what we're doing here is creating a pseudo-random experiment by creating a synthetic control group by selecting people who were *just as likely* to have been in the treatment group but weren't. How do we do this? See the word in that sentence, "likely"? Time to break out the logistic regression. So there are two stages to doing propensity score modeling. The first stage is to use logistic regression to model the probability of each person's likelihood to have *received the treatment*; we then might pair people up so that one person received the treatment and the other didn't, but they had been *equally likely* (or close to equally likely) to have received it. Then we can proceed as we would if we had a random experiment on our hands.

For example, if we wanted to measure the effect of smoking on the probability of lung cancer, we'd have to find people who shared the same probability of *smoking*. We'd collect as many covariates of people as we could (age, whether or not their parents smoked, whether or not their spouses smoked, weight, diet, exercise, hours a week they work, blood test results), and we'd use as an outcome whether or not they smoked. We'd build a logistic regression that predicted the probability of smoking. We'd then use that model to assign to each person the probability, which would be called their propensity score, and then we'd use that to match. Of course we're banking on the fact that we *figured out* and were able to observe *all* the covariates associated with likelihood of smoking, which we're probably not. And that's the inherent difficulty in these methods: we'll never know if we actually adjusted for everything we needed to adjust for. *However*, one of the nice aspects of them is we'll see that when we do adjust for confounders, it can make a big difference in the estimated causal effect.

The details of setting of matching can be slightly more complicated than just paired matching—there are more complex schemes to try to create balance in the synthetic treatment and control group. And there are packages in R that can do it all automatically for you, except you must specify the model that you want to use for matching in the first place to generate the propensity scores, and which variable you want to be the outcome corresponding to the causal effect you are estimating.

What kind of data would we need to measure the causal effect in our dating example? One possibility is to have some third party, a mechanical turk, for example, go through the dating profiles of the women that Frank emails and label the ones that are beautiful. That way we could see to what extent being beautiful is a confounder. This approach is called *stratification* and, as we will see in the next chapter, it can introduce problems as well as fix them.

# Three Pieces of Advice

Ori took a moment to give three pieces of parting advice for best practices when modeling.

First, when estimating causal parameters, it is crucial to understand the data-generating methods and distributions, which will in turn involve gaining some subject matter knowledge. Knowing exactly how the data was generated will also help you ascertain whether the assumptions you make are reasonable.

Second, the first step in a data analysis should always be to take a step back and figure out *what you want to know*. Write it down carefully, and then find and use the tools you've learned to answer those directly. Later on be sure and come back to decide how close you came to answering your original question or questions. Sounds obvious, but you'd be surprised how often people forget to do this.

Finally, don't ignore the necessary data intuition when you make use of algorithms. Just because your method converges, it doesn't mean the results are meaningful. Make sure you've created a reasonable narrative and ways to check its validity.

# CHAPTER 12
# Epidemiology

The contributor for this chapter is David Madigan, professor and chair of statistics at Columbia. Madigan has over 100 publications in such areas as Bayesian statistics, text mining, Monte Carlo methods, pharmacovigilance, and probabilistic graphical models.

## Madigan's Background

Madigan went to college at Trinity College Dublin in 1980, and specialized in math except for his final year, when he took a bunch of stats courses, and learned a bunch about computers: Pascal, operating systems, compilers, artificial intelligence, database theory, and rudimentary computing skills. He then worked in industry for six years, at both an insurance company and a software company, where he specialized in expert systems.

It was a mainframe environment, and he wrote code to price insurance policies using what would now be described as scripting languages. He also learned about graphics by creating a graphic representation of a water treatment system. He learned about controlling graphics cards on PCs, but he still didn't know about data.

Next he got a PhD, also from Trinity College Dublin, and went into academia, and became a tenured professor at the University of Washington. That's when machine learning and data mining started, which he fell in love with: he was program chair of the KDD conference, among other things. He learned C and Java, R, and S+. But he *still* wasn't really working with data yet.

He claims he was still a typical academic statistician: he had computing skills but no idea how to work with a large-scale medical database, 50 different tables of data scattered across different databases with different formats.

In 2000 he worked for AT&T Labs. It was an "extreme academic environment," and he learned perl and did lots of stuff like web scraping. He also learned awk and basic Unix skills.

He then went to an Internet startup where he and his team built a system to deliver real-time graphics on consumer activity.

Since then he's been working in big medical data stuff. He's testified in trials related to medical trials (the word "trial" is used here in two different ways in this sentence), which was eye-opening for him in terms of explaining what you've done: "If you're gonna explain logistic regression to a jury, it's a different kind of a challenge than me standing here tonight." He claims that super simple graphics help.

## Thought Experiment

We now have detailed, longitudinal medical data on tens of millions of patients. What can we do with it?

To be more precise, we have tons of phenomenological data: this is individual, patient-level medical record data. The largest of the databases has records on 80 million people: every prescription drug, every condition ever diagnosed, every hospital or doctor's visit, every lab result, procedures, all timestamped.

But we still do things like we did in the Middle Ages; the vast majority of diagnosis and treatment is done in a doctor's brain. Can we do better? Can we harness these data to do a better job delivering medical care?

This is a hugely important clinical problem, especially as a healthcare insurer. Can we intervene to avoid hospitalizations?

So for example, there was a prize offered on Kaggle, called "Improve Healthcare, Win $3,000,000." It challenged people to accurately predict who is going to go to the hospital next year. However, keep in mind that they've coarsened the data for privacy reasons.

There are a lot of sticky ethical issues surrounding this 80 million person medical record dataset. What nefarious things could we do with this data? Instead of helping people stay well, we could use such models to gouge sick people with huge premiums, or we could drop sick people from insurance altogether.

This is not a modeling question. It's a question of what, as a society, we want to do with our models.

## Modern Academic Statistics

It used to be the case, say 20 years ago, according to Madigan, that academic statisticians would either sit in their offices proving theorems with no data in sight—they wouldn't even know how to run a t-test—or sit around in their offices and dream up a new test, or a new way of dealing with missing data, or something like that, and then they'd look around for a dataset to whack with their new method. In either case, the work of an academic statistician required no domain expertise.

Nowadays things are different. The top stats journals are more deep in terms of application areas, the papers involve deep collaborations with people in social sciences or other applied sciences. Madigan sets an example by engaging with the medical community.

Madigan went on to make a point about the modern machine learning community, which he is or was part of: it's a newish academic field, with conferences and journals, etc., but from his perspective, it's characterized by what statistics was 20 years ago: invent a method, try it on datasets. In terms of domain expertise engagement, it's a step backward instead of forward.

Not to say that statistics are perfect; very few academic statisticians have serious hacking skills, with Madigan's colleague Mark Hansen being an unusual counterexample. In Madigan's opinion, statisticians should not be allowed out of school unless they have such skills.

## Medical Literature and Observational Studies

As you may not be surprised to hear, medical journals are *full* of observational studies. The results of these studies have a profound effect on medical practice, on what doctors prescribe, and on what regulators do.

For example, after reading the paper entitled "Oral bisphosphonates and risk of cancer of oesophagus, stomach, and colorectum: case-control analysis within a UK primary care cohort" (by Jane Green, et al.), Madigan concluded that we see the potential for the very same kind of confounding problem as in the earlier example with aspirin. The conclusion of the paper is that the risk of cancer increased with 10 or more prescriptions of oral bisphosphonates.

It was published on the front page of the *New York Times*, the study was done by a group with no apparent conflict of interest, and the drugs are taken by millions of people. But the results might well be wrong and, indeed, were contradicted by later studies.

There are thousands of examples of this. It's a major problem and people don't even get that it's a problem.

Billions upon billions of dollars are spent doing medical studies, and people's lives depend on the results and the interpretations. We should really know if they work.

## Stratification Does Not Solve the Confounder Problem

The field of epidemiology attempts to adjust for potential confounders. The bad news is that it doesn't work very well. One reason is that the methods most commonly used rely heavily on *stratification*, which means partitioning the cases into subcases and looking at those. So, for example, if they think gender is a confounder, they'd adjust for gender in the estimator—a weighted average is one way of stratifying.

But there's a problem here, too. Stratification could make the underlying estimates of the causal effects go from good to bad, especially when the experiment involves small numbers or when the populations are not truly similar.

For example, say we have the situation shown in Table 12-1. Keep in mind that we cannot actually "see" the two counterfactual middle columns.

*Table 12-1. Aggregated: Both men and women*

|          | Treatment: Drugged | Treatment: Counterfactual | Control: Counterfactual | Control: No Drug |
|----------|--------------------|---------------------------|-------------------------|------------------|
| Y=1      | 30                 | 20                        | 30                      | 20               |
| Y=0      | 70                 | 80                        | 70                      | 80               |
| P(Y=1)   | 0.3                | 0.2                       | 0.3                     | 0.2              |

Here we have 100 people in both the treatment and control groups, and in both the actual and counterfactual cases, we have a causal effect of 0.3 − 0.2 = 0.1, or 10%.

But when we split this up by gender, we might introduce a problem, especially as the numbers get smaller, as seen in Tables 12-2 and 12-3.

*Table 12-2. Stratified: Men*

|  | Treatment: Drugged | Treatment: Counterfactual | Control: Counterfactual | Control: No Drug |
|---|---|---|---|---|
| Y=1 | 15 | 2 | 5 | 5 |
| Y=0 | 35 | 8 | 65 | 15 |
| P(Y=1) | 0.3 | 0.2 | 0.07 | 0.25 |

*Table 12-3. Stratified: Women*

|  | Treatment: Drugged | Treatment: Counterfactual | Control: Counterfactual | Control: No Drug |
|---|---|---|---|---|
| Y=1 | 15 | 18 | 25 | 15 |
| Y=0 | 35 | 72 | 5 | 65 |
| P(Y=1) | 0.3 | 0.2 | 0.83 | 0.1875 |

Our causal estimate for men is 0.3 − 0.25 = 0.05, and for women is 0.3 − 0.1875 = 0.1125. A headline might proclaim that the drug has side effects twice as strong for women as for men.

In other words, stratification doesn't just solve problems. There are no guarantees your estimates will be better if you stratify. In fact, you should have very good evidence that stratification helps before you decide to do it.

## What Do People Do About Confounding Things in Practice?

In spite of the raised objections, experts in this field essentially use stratification as a major method to working through studies. They deal with confounding variables, or rather variables they deem potentially confounding, by stratifying with respect to them or make other sorts of model-based adjustments, such as propensity score matching, for example. So if taking aspirin is believed to be a potentially confounding factor, they adjust or stratify with respect to it.

For example, with this study (*http://goo.gl/3VgRi0*), which studied the risk of venous thromboembolism from the use of certain kinds of oral contraceptives, the researchers chose certain confounders to worry about and concluded the following:

> After adjustment for length of use, users of oral contraceptives with desogestrel, gestodene, or drospirenone were at least at twice the risk of venous thromboembolism compared with users of oral contraceptives with levonorgestrel.

This report was featured on ABC, and was a big deal. But wouldn't you worry about confounding issues like *aspirin* here? How do you choose which confounders to worry about? Or, wouldn't you worry that the physicians who are prescribing them act different in different situations, leading to different prescriptions? For example, might they give the newer one to people at higher risk of clotting?

Another study came out about this same question and came to a different conclusion, using different confounders. The researchers adjusted for a history of clots, which makes sense when you think about it. Altogether we can view this as an illustration of how, depending on how one chooses to adjust for things, the outputs can vary wildly. It's starting to seem like a hit or miss methodology.

Another example is a study on oral bisphosphonates, where they adjusted for smoking, alcohol, and BMI. How did they choose those variables? In fact, there are hundreds of examples where two teams made radically different choices on parallel studies.

Madigan and some coauthors tested this by giving a bunch of epidemiologists the job to design five studies at a high level. There was a low-level consistency. However, an additional problem is that luminaries of the field hear this and claim that they know the "right" way to choose the confounders.

## Is There a Better Way?

Madigan and his coauthors examined 50 studies, each of which corresponds to a drug and outcome pair (e.g., antibiotics with GI bleeding). They ran about 5,000 analyses for every pair—namely, every epistudy imaginable—and they did this all on nine different databases.

For example, they fixed the drug to be ACE inhibitors and the outcome to be swelling of the heart. They ran the same analysis on the nine different standard databases, the smallest of which has records of

4,000,000 patients, and the largest of which has records of 80,000,000 patients.

In this one case, for one database, the drug triples the risk of heart swelling; but for another database, it seems to have a six-fold increase of risk. That's one of the best examples, though, because at least it's *always bad news*, which means it's consistent.

On the other hand, for 20 of the 50 pairs, you can go from statistically significant in one direction to the other direction depending on the database you pick. In other words, you can get *whatever you want*. Figure 12-1 shows a picture, where the heart swelling example is at the top.

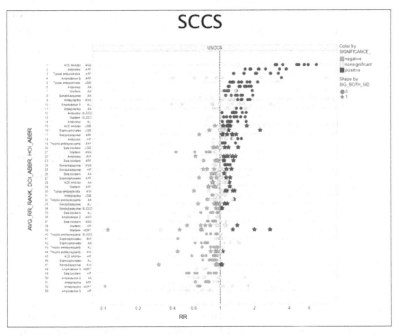

*Figure 12-1. Self-controlled case series*

The choice of database is rarely discussed in published epidemiology papers.

Next they did an even more extensive test, where they essentially tried everything. In other words, every time there was a decision to be made, they did it both ways. The kinds of decisions they tweaked were as follows: which database you tested on, the confounders you accounted for, and the window of time you cared about examining, which refers to the situation where a patient has a heart attack a week or a month after discontinuing a treatment and whether that is counted in the study.

What they saw was that *almost all the studies can get either side depending on the choices.*

Let's get back to oral bisphosphonates. A certain study (*http://1.usa.gov/16UfNjZ*) concluded that they cause esophageal cancer, but two weeks later, JAMA published a paper (*http://1.usa.gov/1hi2kbj*) on the same issue that concluded they are *not* associated with elevated risk of esophageal cancer. And they were even using the *same* database. This is not so surprising now for us.

# Research Experiment (Observational Medical Outcomes Partnership)

To address the issues directly, or at least bring to light the limitations of current methods and results, Madigan has worked as a principal investigator on the OMOP (*http://omop.org/*) research program, making significant contributions to the project's methodological work including the development, implementation, and analysis of a variety of statistical methods applied to various observational databases.

## About OMOP, from Its Website

In 2007, recognizing that the increased use of electronic health records (EHR) and availability of other large sets of marketplace health data provided new learning opportunities, Congress directed the FDA to create a new drug surveillance program to more aggressively identify potential safety issues. The FDA launched several initiatives to achieve that goal, including the well-known Sentinel program to create a nationwide data network.

In partnership with PhRMA and the FDA, the Foundation for the National Institutes of Health launched the Observational Medical Outcomes Partnership (OMOP), a public-private partnership. This interdisciplinary research group has tackled a surprisingly difficult task that is critical to the research community's broader aims: identifying the most reliable methods for analyzing huge volumes of data drawn from heterogeneous sources.

Employing a variety of approaches from the fields of epidemiology, statistics, computer science, and elsewhere, OMOP seeks to answer a critical challenge: what can medical researchers learn from assessing these new health databases, could a single approach be applied to multiple diseases, and could their findings be proven? Success would mean the opportunity for the medical research community to do more studies in less time, using fewer resources and achieving more consistent results. In the end, it would mean a better system for monitoring drugs, devices, and procedures so that the healthcare community can reliably identify risks and opportunities to improve patient care.

Madigan and his colleagues took 10 large medical databases, consisting of a mixture of claims from insurance companies and electronic health records (EHR), covering records of 200 million people in all. This is Big Data unless you talk to an astronomer.

They mapped the data to a common data model and then they implemented every method used in observational studies in healthcare. Altogether they covered 14 commonly used epidemiology designs adapted for longitudinal data. They automated everything in sight. Moreover, there were about 5,000 different "settings" on the 14 methods.

The idea was to see how well the current methods do on predicting *things we actually already know*.

To locate things they know, they took 10 old drug classes: ACE inhibitors, beta blockers, warfarin, etc., and 10 outcomes of interest: renal failure, hospitalization, bleeding, etc.

For some of these, the results are known. So, for example, warfarin is a blood thinner and definitely causes bleeding. There were nine such known bad effects.

There were also 44 known "negative" cases, where we are super confident there's just no harm in taking these drugs, at least for these outcomes.

The basic experiment was this: run 5,000 commonly used epidemiological analyses using all 10 databases. How well do they do at discriminating between reds and blues? Kind of like a spam filter test, where one has training emails that are known spam, and one wants to know how well the model does at detecting spam when it comes through.

Each of the models output the same thing: a relative risk (RR) [measured by the causal effect estimate we talked about previously] and an error.

Theirs was an attempt to empirically evaluate how well epidemiology works, kind of the quantitative version of John Ioannidis's work (*http:// stanford.io/15LfJDL*).

### Why Hasn't This Been Done Before?
There's conflict of interest for epidemiology—why would they want to prove their methods don't work? Also, it's expensive: it cost 25 million dollars, which of course pales in comparison to the money being put into these studies.

They bought all the data, made the methods work automatically, and did a bunch of calculations in the Amazon cloud. The code is open source. In the second version, they zeroed in on four particular outcomes and built the $25,000,000 so-called ROC curve shown in Figure 12-2.

To understand this graph, we need to define a *threshold*, which we can start with at 2. This means that if the relative risk is estimated to be above 2, we call it a "bad effect"; otherwise we call it a "good effect." The choice of threshold will, of course, matter.

If it's high, say 10, then you'll never see a 10, so everything will be considered a good effect. Moreover, these are old drugs and wouldn't be on the market. This means your sensitivity will be low, and you won't find any real problem. That's bad! You should find, for example, that warfarin causes bleeding.

There's of course good news too, with low sensitivity, namely a zero false-positive rate.

What if you set the threshold really low, at −10? Then everything's bad, and you have a 100% sensitivity but very high false-positive rate.

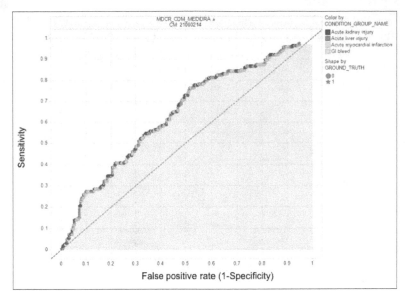

Figure 12-2. The $25,000,000 ROC curve

As you vary the threshold from very low to very high, you sweep out a curve in terms of sensitivity and false-positive rate, and that's the curve we see in the figure. There is a threshold (say, 1.8) for which your false positive rate is 30% and your sensitivity is 50%.

 This graph is seriously problematic if you're the FDA. A 30% false-positive rate is not within the parameters that the FDA considers acceptable.

The overall "goodness" of such a curve is usually measured as the area under the curve (AUC): you want it to be one, and if your curve lies on diagonal, the area is 0.5. This is tantamount to guessing randomly. So if your area under the curve is less than 0.5, it means your model is perverse.

The AUC in the preceding figure is 0.64. Moreover, of the 5,000 analyses run by the research team (which included David Madigan), this is the single best analysis.

But note: this is the best *if you can only use the same method for everything*. In that case this is as good as it gets, and it's not that much better than guessing.

One the other hand, no epidemiologist would do that. So what they did next was to *specialize the analysis to the database and the outcome*. And they got better results: for the Medicare database, and for acute kidney injury, their optimal model gives them an AUC of 0.92 as shown in Figure 12-3. They can achieve 80% sensitivity with a 10% false-positive rate.

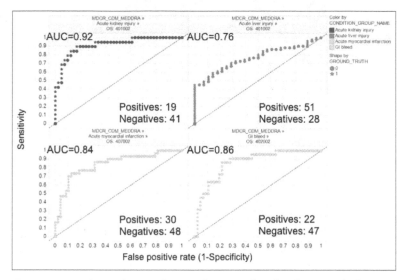

*Figure 12-3. Specializing the analysis to the database and the outcome with better results*

They did this using a cross-validation method. Different databases have different methods attached to them. One winning method is called "OS," which compares within a given patient's history (so compares times when the patient was on drugs versus when they weren't). This is not widely used now.

The epidemiologists in general don't believe the results of this study.

If you go to *http://elmo.omop.org*, you can see the AUC for a given database and a given method. The data used in this study was current in mid-2010. To update this, you'd have to get the latest version of the database, and rerun the analysis. Things might have changed.

# Closing Thought Experiment

In the study, 5,000 different analyses were run. Is there a good way of combining them to do better? How about incorporating weighted averages or voting methods across different strategies? The code is publicly available and might make a great PhD thesis.

# CHAPTER 13
# Lessons Learned from Data Competitions: Data Leakage and Model Evaluation

The contributor for this chapter is Claudia Perlich. Claudia has been the Chief Scientist at Media 6 Degrees (M6D) (*http://m6d.com*) for the past few years. Before that she was in the data analytics group at the IBM center that developed Watson, the computer that won Jeopardy! (although she didn't work on that project). Claudia holds a master's in computer science, and got her PhD in information systems at NYU. She now teaches a class to business students on data science, where she addresses how to assess data science work and how to manage data scientists.

Claudia is also a famously successful data mining competition winner. She won the KDD Cup in 2003, 2007, 2008, and 2009, the ILP Challenge in 2005, the INFORMS Challenge in 2008, and the Kaggle HIV competition in 2010.

More recently she's turned toward being a data mining competition organizer, first for the INFORMS Challenge in 2009, and then for the Heritage Health Prize in 2011. Claudia claims to be retired from competition. Fortunately for the class, she provided some great insights into what can be learned from data competitions. From the many competitions she's done, she's learned quite a bit in particular about data leakage, and how to evaluate the models she comes up with for the competitions.

# Claudia's Data Scientist Profile

Claudia started by asking what people's reference point might be to evaluate where they stand with their own data science profile (hers is shown in Table 13-1. Referring to the data scientist profile from Chapter 1, she said, "There is one skill that you do not have here and that is the most important and the hardest to describe: Data." She knows some of the world's best mathematicians, machine language experts, statisticians, etc. Does she calibrate herself toward what is possible (the experts) or just relative to the average person in her field, or just an average person?

*Table 13-1. Claudia's data science profile*

|  | passable | strong | solid | comment |
|---|---|---|---|---|
| Visualization | x |  |  | I can do it but I do not believe in visualization |
| Computer Science | x |  |  | I have 2 Masters Degrees in CS. I can hack, not production code. |
| Math | x |  |  | That was a long time ago |
| Stats |  | x |  | Little formal training, a lot of stuff picked up on the way and good intuition |
| Machine Learning |  |  | x |  |
| Domain |  |  |  | You are asking the wrong question… |
| Presentation |  |  | x |  |
| Data |  |  | x |  |

# The Life of a Chief Data Scientist

Historically, Claudia has spent her time on predictive modeling, including data mining competitions, writing papers for publications and conferences like KDD and journals, giving talks, writing patents, teaching, and digging around data (her favorite part). She likes to understand something about the world by looking directly at the data.

Claudia's skill set includes 15 years working with data, where she's developed data intuition by delving into the *data generating process*, a crucial piece of the puzzle. She's given a lot of time and thought to the evaluation process and developing model intuition.

Claudia's primary skills are data manipulation using tools like Unix, sed, awk, Perl, and SQL. She models using various methods, including logistic regression, k-nearest neighbors, and, importantly, she spends a bunch of time setting things up well. She spends about 40% of her time as "contributor," which means doing stuff directly with data; 40% of her time as "ambassador," which means writing stuff, and giving talks, mostly external communication to represent M6D; and 20% of her time in "leadership" of her data group.

## On Being a Female Data Scientist

Being a woman works well in the field of data science, where intuition is useful and is regularly applied. One's nose gets so well developed by now that one can *smell* it when something is wrong, although this is not the same thing as being able to prove something algorithmically. Also, people typically remember women, even when women don't remember them. It has worked in her favor, Claudia says, which she's happy to admit. But then again, she is where she is fundamentally because she's good.

Being in academia, Claudia has quite a bit of experience with the process of publishing her work in journals and the like. She discussed whether papers submitted for journals and/or conferences are blind to gender. For some time, it was typically double-blind, but now it's more likely to be one-sided. Moreover, there was a 2003 paper written by Shawndra Hill and Foster Provost that showed you can guess who wrote a paper with 40% accuracy just by knowing the citations, and even more if the author had more publications. Hopefully people don't actually *use* such models when they referee, but in any case, that means making things "blind" doesn't necessarily help. More recently the names are included, and hopefully this doesn't make things too biased. Claudia admits to being slightly biased herself toward institutions—in her experience, certain institutions prepare better work.

# Data Mining Competitions

Claudia drew a distinction between different types of data mining competitions. The first is the "sterile" kind, where you're given a clean, prepared data matrix, a standard error measure, and features that are often anonymized. This is a pure machine learning problem.

Examples of this first kind are KDD Cup 2009 and the Netflix Prize, and many of the Kaggle competitions. In such competitions, your approach would emphasize algorithms and computation. The winner would probably have heavy machines and huge modeling ensembles.

# KDD Cups

All the KDD Cups, with their tasks and corresponding datasets, can be found at *http://www.kdd.org/kddcup/index.php*. Here's a list:

- KDD Cup 2010: Student performance evaluation
- KDD Cup 2009: Customer relationship prediction
- KDD Cup 2008: Breast cancer
- KDD Cup 2007: Consumer recommendations
- KDD Cup 2006: Pulmonary embolisms detection from image data
- KDD Cup 2005: Internet user search query categorization
- KDD Cup 2004: Particle physics; plus protein homology prediction
- KDD Cup 2003: Network mining and usage log analysis
- KDD Cup 2002: BioMed document; plus gene role classification
- KDD Cup 2001: Molecular bioactivity; plus protein locale prediction
- KDD Cup 2000: Online retailer website clickstream analysis
- KDD Cup 1999: Computer network intrusion detection
- KDD Cup 1998: Direct marketing for profit optimization
- KDD Cup 1997: Direct marketing for lift curve optimization

On the other hand, you have the "real world" kind of data mining competition, where you're handed raw data (which is often in lots of different tables and not easily joined), you set up the model yourself, and come up with task-specific evaluations. This kind of competition simulates real life more closely, which goes back to Rachel's thought experiment earlier in this book about how to simulate the chaotic experience of being a data scientist in the classroom. You need practice dealing with messiness.

Examples of this second kind are KDD cup 2007, 2008, and 2010. If you're in this kind of competition, your approach would involve understanding the domain, analyzing the data, and building the model. The winner might be the person who best understands how to tailor the model to the actual question.

Claudia prefers the second kind, because it's closer to what you do in real life.

# How to Be a Good Modeler

Claudia claims that data and domain understanding are the single most important skills you need as a data scientist. At the same time, this can't really be taught—it can only be *cultivated*.

A few lessons learned about data mining competitions that Claudia thinks are overlooked in academics:

*Leakage*
> The contestants' best friend and the organizer and practitioners' worst nightmare. There's always something wrong with the data, and Claudia has made an artform of figuring out how the people preparing the competition got lazy or sloppy with the data.

*Real-life performance measures*
> Adapting learning beyond standard modeling evaluation measures like mean squared error (MSE), misclassification rate, or area under the curve (AUC). For example, profit would be an example of a real-life performance measure.

*Feature construction/transformation*
> Real data is rarely flat (i.e., given to you in a beautiful matrix) and good, practical solutions for this problem remain a challenge.

# Data Leakage

In a KDD 2011 paper that Claudia coauthored called "Leakage in Data Mining: Formulation, Detection, and Avoidance", she, Shachar Kaufman, and Saharon Rosset point to another author, Dorian Pyle, who has written numerous articles and papers on data preparation in data mining, where he refers to a phenomenon that he calls anachronisms (something that is out of place in time), and says that "too good to be true" performance is "a dead giveaway" of its existence. Claudia and her coauthors call this phenomenon "data leakage" in the context of predictive modeling. Pyle suggests turning to exploratory data analysis in order to find and eliminate leakage sources. Claudia and her coauthors sought a rigorous methodology to deal with leakage.

Leakage refers to information or data that helps you predict something, and the fact that you are using this information to predict isn't fair. It's a huge problem in modeling, and not just for competitions. Oftentimes it's an artifact of reversing cause and effect. Let's walk through a few examples to get a feel for how this might happen.

## Market Predictions

There was a competition where you needed to predict S&P in terms of whether it would go up or go down. The winning entry had an AUC (area under the ROC curve) of 0.999 out of 1. Because stock markets are pretty close to random, either someone's very rich or there's something wrong. (Hint: there's something wrong.)

In the good old days you could win competitions this way, by finding the leakage. It's not clear in this case what the leakage was, and you'd only know if you started digging into the data and finding some piece of information in the dataset that was highly predictive of S & P, but that would *not* be available to you in real time when predicting S & P. We bring this example up because the very fact that they had such a high AUC means that their model must have been relying on leakage, and it would not work if implemented.

## Amazon Case Study: Big Spenders

The target of this competition was to predict customers who will be likely to spend a lot of money by using historical purchase data. The data consisted of transaction data in different categories. But a winning model identified that "Free Shipping = True" was an excellent predictor of spending a lot. Now notice, you only get offered free shipping *after* you've spent a certain amount of money, say above $50.

What happened here? The point is that free shipping is an *effect* of big spending. But it's not a good way to model big spending, because in particular, it doesn't work for new customers or for the future. Note: timestamps are weak here. The data that included "Free Shipping = True" was simultaneous with the sale, which is a no-no. You need to only use data from beforehand to predict the future. The difficulty is that this information about free shipping appeared in the collected data, and so it has to be manually removed, which requires consideration and *understanding* the data on the part of the model builder. If you weren't thinking about leakage, you could just throw the free shipping variable into your model and see it predicted well. But, then when you actually went to implement the model in production, you wouldn't know that the person was *about* to get free shipping.

# A Jewelry Sampling Problem

Again for an online retailer, this time the target was predicting customers who buy jewelry. The data consisted of transactions for different categories. A very successful model simply noted that if sum(revenue) = 0, then it predicted jewelry customers very well.

What happened here? The people preparing the data for the competition removed jewelry purchases, but only included people who bought something in the first place. So people who had sum(revenue) = 0 were people who only bought jewelry. The fact that only people who bought something got into the dataset is weird: in particular, you wouldn't be able to use this on customers before they finished their purchase. So the model wasn't being trained on the right data to make the model useful. This is a sampling problem, and it's common.

---

## Warning: Sampling Users

As mentioned, in this case it's weird to only condition the analysis on the set of people who have already bought something. Do you really want to condition your analysis on *only* people who bought something or *all* people who came to your site? More generally with user-level data, if you're not careful, you can make fairly simple, but serious sampling mistakes if you don't think it through. For example, say you're planning to analyze a dataset from one day of user traffic on your site. If you do this, you are *oversampling* users who come frequently.

Think of it this way with a toy example: suppose you have 80 users. Say that 10 of them come every day, and the others only come once a week. Suppose they're spread out evenly over 7 days of the week. So on any given day you see 20 users. So you pick a day. You look at those 20 users—10 of them are the ones who come every day, the other 10 come once a week. What's happening is that you're oversampling the users who come every day. Their behavior on your site might be totally different than other users, and they're representing 50% of your dataset even though they only represent 12.5% of your user base.

---

# IBM Customer Targeting

At IBM, the target was to predict companies that would be willing to buy "websphere" solutions. The data was transaction data and crawled

---

potential company websites. The winning model showed that if the term "websphere" appeared on the company's website, then it was a great candidate for the product. What happened? Remember, when considering a potential customer, by definition that company wouldn't have bought websphere *yet* (otherwise IBM wouldn't be trying to sell to it); therefore no *potential* customer would have websphere on its site, so it's not a predictor at all. If IBM could go back in time to see a snapshot of the historical Web as a source of data before the "websphere" solution product existed, then that data would make sense as a predictor. But using today's data unfortunately contains the leaked information that they ended up buying websphere. You can't crawl the *historical* Web, just today's Web.

Seem like a silly, obvious mistake not to make? Maybe. But it's the sort of thing that happens, and you can't anticipate this kind of thing until you start digging into the data and really understand what features and predictors *mean*. Just think, if this happened with something "obvious," it means that more careful thought and digging needs to go on to figure out the less obvious cases. Also, it's an example of something maybe we haven't emphasized enough yet in the book. Doing simple sanity checking to make sure things are what you think they are can sometimes get you much further in the end than web scraping and a big fancy machine learning algorithm. It may not seem cool and sexy, but it's smart and good practice. People might not invite you to a meetup to talk about it. It may not be publishable research, but at least it's legitimate and solid work. (Though then again, because of this stuff, Claudia won tons of contests and gets invited to meetups all the time. So we take that back. No, we don't. The point is, do good work, the rest will follow. Meetups and fame aren't goals unto themselves. The pursuit of the truth is.)

## Breast Cancer Detection

You're trying to study who has breast cancer. Take a look at Figure 13-1. The patient ID, which seems innocent, actually has predictive power. What happened?

In Figure 13-1, red means cancerous, green means not; it's plotted by patient ID. We see three or four distinct buckets of patient identifiers. It's very predictive depending on the bucket. This is probably a consequence of using multiple databases corresponding to different cancer centers, some of which take on sicker patients—by definition patients who get assigned to that center are more likely to be sick.

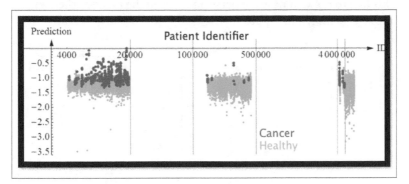

*Figure 13-1. Patients ordered by patient identifier; red means cancerous, green means not*

This situation led to an interesting discussion in the classroom:

> *Student*: For the purposes of the contest, they should have renumbered the patients and randomized.
>
> *Claudia*: Would that solve the problem? There could be other things in common as well.
>
> *Another student*: The important issue could be to see the extent to which we can figure out which dataset a given patient came from based on things besides their ID.
>
> *Claudia*: Think about this: what do we want these models for in the first place? How well can you *really* predict cancer?

Given a new patient, what would *you* do? If the new patient is in a fifth bin in terms of patient ID, then you wouldn't want to use the identifier model. But if it's still in this scheme, then maybe that really is the best approach.

This discussion brings us back to the fundamental problem: we need to know what the purpose of the model is and how it is going to be used in order to decide how to build it, and whether it's working.

## Pneumonia Prediction

During an INFORMS competition on pneumonia predictions in hospital records—where the goal was to predict whether a patient has pneumonia—a logistic regression that included the number of diagnosis codes as a numeric feature (AUC of 0.80) didn't do as well as the one that included it as a categorical feature (0.90). What happened?

This had to do with how the person prepared the data for the competition, as depicted in Figure 13-2.

| icd1x | icd2x | icd3x | icd4x |
|-------|-------|-------|-------|
| 786 | 285 | 459 | -1 |
| 401 | 486 | -1 | -1 |
| 401 | 486◄······780 | -1 | -1 |
| 599 | -1 | -1 | -1 |
| V22 | 650 | -1 | -1 |
| V56 | 492 | 586 | -1 |
| 786 | 493 | 285 | 459 |

→

| icd1x | icd2x | icd3x | icd4x |
|-------|-------|-------|-------|
| 786 | 285 | 459 | -1 |
| 401 | -1 | -1 | -1 |
| 401 | 780 | -1 | -1 |
| 599 | -1 | -1 | -1 |
| V22 | 650 | -1 | -1 |
| V56 | 492 | 586 | -1 |
| 786 | 493 | 285 | 459 |

*Figure 13-2. How data preparation was done for the INFORMS competition*

The diagnosis code for pneumonia was 486. So the preparer removed that (and replaced it with a "–1") if it showed up in the record (rows are different patients; columns are different diagnoses; there is a maximum of four diagnoses; "–1" means there's nothing for that entry).

Moreover, to avoid telling holes in the data, the preparer moved the other diagnoses to the left if necessary, so that only "–1"s were on the right.

There are two problems with this:

- If the row has only "–1"s, then you know it started out with only pneumonia.
- If the row has no "–1"s, you know there's no pneumonia (unless there are actually five diagnoses, but that's less common).

This alone was enough information to win the competition.

**Leakage Happens**

Winning a competition on leakage is easier than building good models. But even if you don't explicitly understand and game the leakage, your model will do it for you. Either way, leakage is a huge problem with data mining contests in general.

# How to Avoid Leakage

The message here is not about how to win predictive modeling competitions. The reality is that as a data scientist, you're at risk of producing a data leakage situation any time you prepare, clean your data, impute missing values, remove outliers, etc. You might be distorting the data in the process of preparing it to the point that you'll build a model that works well on your "clean" dataset, but will totally suck when applied in the real-world situation where you actually want to apply it. Claudia gave us some very specific advice to avoid leakage. First, you need a strict temporal cutoff: remove all information just prior to the event of interest. For example, stuff you know *before a patient is admitted*. There has to be a timestamp on every entry that corresponds to the time you learned that information, not the time it occurred. Removing columns and rows from your data is asking for trouble, specifically in the form of inconsistencies that can be teased out. The best practice is to start from scratch with unfiltered, raw data after careful consideration. Finally, you need to know how the data was created!

Claudia and her coauthors describe in the paper referenced earlier a suggested methodology for avoiding leakage as a two-stage process of tagging every observation with legitimacy tags during collection and then observing what they call a learn-predict separation.

# Evaluating Models

How do you know that your model is any good? We've gone through this already in some previous chapters, but it's always good to hear this again from a master.

With powerful algorithms searching for patterns of models, there is a serious danger of overfitting. It's a difficult concept, but the general idea is that "if you look hard enough, you'll find something," even if it does not generalize beyond the particular training data.

To avoid overfitting, we cross-validate and cut down on the complexity of the model to begin with. Here's a standard picture in Figure 13-3 (although keep in mind we generally work in high dimensional space and don't have a pretty picture to look at).

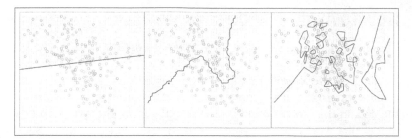

*Figure 13-3. This classic image from Hastie and Tibshirani's Elements of Statistical Learning (http://stanford.io/17szrYz) (Springer-Verlag) shows fitting linear regression to a binary response, fitting 15-nearest neighbors, and fitting 1-nearest neighbors all on the same dataset*

The picture on the left is underfit, in the middle it's good, and on the right it's overfit.

The model you use matters when it concerns overfitting, as shown in Figure 13-4.

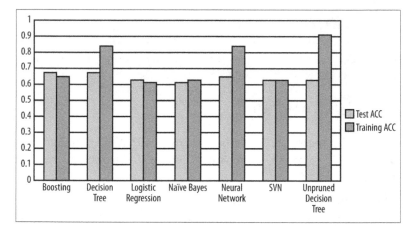

*Figure 13-4. The model you use matters!*

Looking at Figure 13-4, unpruned decision trees are the overfitting-est (we just made that word up). This is a well-known problem with unpruned decision trees, which is why people use pruned decision trees.

## Accuracy: Meh

One of the model evaluation metrics we've discussed in this book is accuracy as a means to evaluate classification problems, and in particular binary classification problems. Claudia dismisses accuracy as a bad evaluation method. What's wrong with accuracy? It's inappropriate for regression obviously, but even for classification, if the vast majority of binary outcomes are 1, then a stupid model can be accurate but not good ("guess it's always 1"), and a better model might have lower accuracy.

## Probabilities Matter, Not 0s and 1s

Nobody makes decisions on the binary outcomes *themselves*. You want to know the *probability* you'll get breast cancer; you don't want to be told yes or no. It's much more information to know a probability. People *care* about probabilities.

So how does Claudia think evaluation should be handled? She's a proponent of evaluating the ranking and the calibration separately. To evaluate the ranking, we use the ROC curve and calculate the area under it, which typically ranges from 0.5 to 1.0. This is independent of scaling and calibration. Figure 13-5 shows an example of how to draw an ROC curve.

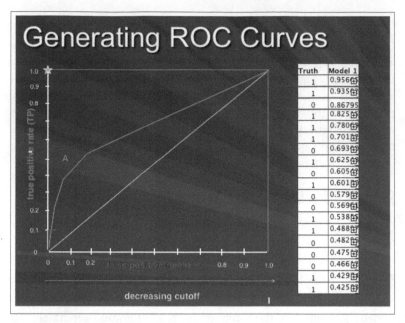

Figure 13-5. *An example of how to draw an ROC curve*

Sometimes to measure rankings, people draw the so-called lift curve shown in Figure 13-6.

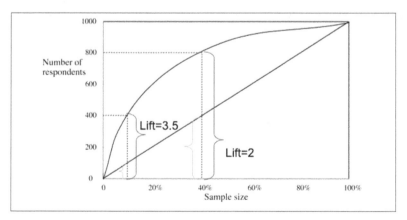

Figure 13-6. *The so-called lift curve*

The key here is that the lift is calculated *with respect to a baseline*. You draw it at a given point, say 10%, by imagining that 10% of people are shown ads, and seeing how many people click versus if you randomly showed 10% of people ads. A lift of 3 means it's 3 times better.

How do you measure calibration? Are the probabilities accurate? If the model says probability of 0.57 that I have cancer, how do I know if it's really 0.57? We can't measure this directly. We can only bucket those predictions and then aggregately compare those in that prediction bucket (say 0.50–0.55) to the actual results for that bucket.

For example, take a look at Figure 13-7, which shows what you get when your model is an unpruned decision tree, where the blue diamonds are buckets.

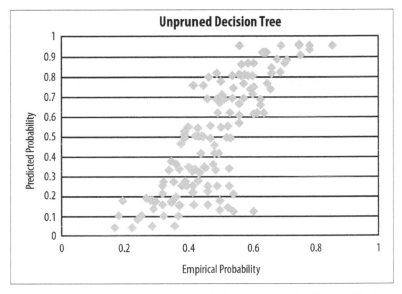

*Figure 13-7. A way to measure calibration is to bucket predictions and plot predicted probability versus empirical probability for each bucket—here, we do this for an unpruned decision tree*

Blue diamonds are buckets of people, say. The x-axis is the empirical, observed fraction of people in that bucket who have cancer, as an example, and the y-axis is the average predicted value for that set of people by the unpruned decision tree. This shows that decision trees don't generally do well with calibration.

A good model would show buckets right along the $x = y$ curve, but here we're seeing that the predictions were much more extreme than the actual probabilities. Why does this pattern happen for decision trees?

Claudia says that this is because *trees optimize purity*: it seeks out pockets that have only positives or negatives. Therefore its predictions are more extreme than reality. This is generally true about decision trees: they do not generally perform well with respect to calibration.

Logistic regression looks better when you test calibration, which is typical, as shown in Figure 13-8.

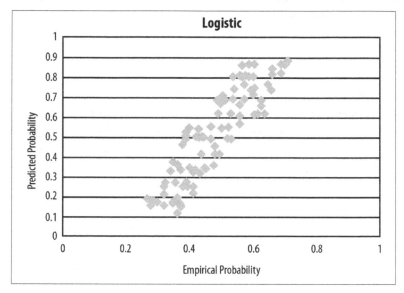

*Figure 13-8. Testing calibration for logistic regression*

Again, blue diamonds are buckets of people. This shows that logistic regression does better with respect to calibration.

# Choosing an Algorithm

This is not a trivial question and, in particular, tests on smaller datasets may steer you wrong, because as you increase the sample size, the best algorithm might vary. Often decision trees perform very well, but only if there's enough data.

In general, you need to choose your algorithm depending on the size and nature of your dataset, and you need to choose your evaluation method based partly on your data and partly on what you wish to be good at. Sum of squared error is the maximum likelihood loss function if your data can be assumed to be normal, but if you want to estimate the median, then use absolute errors. If you want to estimate a quantile, then minimize the weighted absolute error.

We worked on a competition about predicting the number of ratings a movie will get in the next year, and we assumed Poisson distributions. In this case, our evaluation method didn't involve minimizing the sum of squared errors, but rather something else that we found in the literature specific to the Poisson distribution, which depends on the single parameter $\lambda$. So sometimes you need to dig around in the literature to find an evaluation metric that makes sense for your situation.

# A Final Example

Let's put some of this together.

Say you want to raise money for a charity. If you send a letter to every person in the mailing list, you raise about $9,000. You'd like to save money and only send a letter to people who are likely to give—only about 5% of people generally give. How can you figure out who those people are?

If you use a (somewhat pruned, as is standard) decision tree, you get $0 profit: it never finds a leaf with majority positives.

If you use a neural network, you still make only $7,500, even if you only send a letter in the case where you expect the return to be higher than the cost.

Let's break the problem down more. People getting the letter make two decisions: first, they decide whether or not to give, then they decide how much to give. You can model those two decisions separately, using:

$$E(\$|person) = P(response = `yes`|person) \cdot$$
$$E(\$|response = `yes`, person)$$

Note that you need the first model to be well-calibrated because you really care about the number, not just the ranking. So you can try logistic regression for the first half. For the second part, you train with special examples where there *are* donations.

Altogether, this decomposed model makes a profit of $15,000. The decomposition made it easier for the model to pick up the signals. Note that with infinite data, all would have been good, and you wouldn't have needed to decompose. But you work with what you got.

Moreover, you are multiplying errors with this approach, which could be a problem if you have a reason to believe that those errors are correlated.

## Parting Thoughts

According to Claudia, humans are not meant to understand data. Data is outside of our sensory systems, and there are very few people who have a near-sensory connection to numbers. We are instead designed to understand language.

We are also not meant to understand uncertainty: we have all kinds of biases that prevent this from happening that are well documented. Hence, modeling people in the future is intrinsically harder than figuring out how to label things that have already happened.

Even so, we do our best, and this is through careful data generation, meticulous consideration of what our problem is, making sure we model it with data close to how it will be used, making sure we are optimizing to what we actually desire, and doing our homework to learn which algorithms fit which tasks.

# CHAPTER 14

# Data Engineering: MapReduce, Pregel, and Hadoop

We have two contributors to this chapter, David Crawshaw and Josh Wills. Rachel worked with both of them at Google on the Google+ data science team, though the two of them never actually worked *together* because Josh Wills left to go to Cloudera and David Crawshaw replaced him in the role of tech lead. We can call them "data engineers," although that term might be as problematic (or potentially overloaded) or ambiguous as "data scientist"—but suffice it to say that they've both worked as software engineers and dealt with massive amounts of data. If we look at the data science process from Chapter 2, Josh and David were responsible at Google for collecting data (frontend and backend logging), building the massive data pipelines to store and munge the data, and building up the engineering infrastructure to support analysis, dashboards, analytics, A/B testing, and more broadly, data science.

In this chapter we'll hear firsthand from Google engineers about MapReduce, which was developed at Google, and then open source versions were created elsewhere. MapReduce is an algorithm and framework for dealing with massive amounts of data that has recently become popular in industry. The goal of this chapter is to clear up some of the mysteriousness surrounding MapReduce. It's become such a buzzword, and many data scientist job openings are advertised as saying "must know Hadoop" (the open source implementation of MapReduce). We suspect these ads are written by HR departments who don't really understand what MapReduce is good for and the fact that not *all* data science problems *require* MapReduce. But as it's become such a data science term, we want to explain clearly what it is, and where it came from. You should know what it is, but you may not have to use it—or you might, depending on your job.

---

## Do You Need to Know MapReduce to Be a Data Scientist?

A fun game would be to go to a conference, count how many times people say the word "MapReduce," then ask them to actually explain it, and count how many can. We suspect not many can. Even we need to call in the experts who have spent countless hours working with it extensively. At Google, Rachel *did* write code using Sawzall, a programming language that had the MapReduce framework underlying its logic to process and munge the data to get it into shape for analysis and prototyping. For that matter, Cathy used an open source version of Sawzall called Pig when she worked as a data scientist at Intent Media—specifically, she used Pig in conjunction with Python in the Mortar Data framework. So we did indirectly use MapReduce, and we did understand it, but not to the extent these guys did building the underlying guts of the system.

---

Another reason to discuss MapReduce is that it illustrates the types of algorithms used to tackle engineering and infrastructure issues when we have lots of data. This is the third category of algorithms we brought up in Chapter 3 (the other two being machine learning algorithms and optimization algorithms). As a point of comparison, given that algorithmic thinking may be new to you, we'll also describe another data engineering algorithm and framework, Pregel, which enables large-scale graph computation (it was also developed at Google and open sourced).

# About David Crawshaw

David Crawshaw is a Software Engineer at Google who once accidentally deleted 10 petabytes of data with a bad shell script. Luckily, he had a backup. David was trained as a mathematician, worked on Google+ in California with Rachel, and now builds infrastructure for better understanding search. He recently moved from San Francisco to New York.

David came to talk to us about MapReduce and how to deal with too much data. Before we dive in to that, let's prime things with a thought experiment.

# Thought Experiment

How do we think about access to medical records and privacy?

On the one hand, there are very serious privacy issues when it comes to health records—we don't want just anyone to be able to access someone's medical history. On the other hand, certain kinds of access can save lives.

By some estimates, one or two patients died per week in a certain smallish town because of the lack of information flow between the hospital's emergency room and the nearby mental health clinic.[1] In other words, if the records had been easier to match, they'd have been able to save more lives. On the other hand, if it had been easy to match records, other breaches of confidence might also have occurred. Of course it's hard to know exactly how many lives are at stake, but it's nontrivial.

This brings up some natural questions: What is the appropriate amount of privacy in health? Who should have access to your medical records and under what circumstances?

We can assume we think privacy is a generally good thing. For example, to be an atheist is punishable by death in some places, so it's better to be private about stuff in those conditions. But privacy takes lives too, as we see from this story of emergency room deaths.

We can look to other examples, as well. There are many egregious violations happening in law enforcement, where you have large databases of license plates, say, and people who have access can abuse that information. Arguably, though, in this case it's a *human* problem, not a technical problem.

It also can be posed as a philosophical problem: to what extent are we allowed to make decisions on behalf of other people?

---

1. Andrew Gelman thinks this parable is unlikely, and he wrote up a response, which you can read at *http://andrewgelman.com/2014/01/24/parables-vs-data*.

There's also a question of incentives. We might cure cancer faster with more medical data, but we can't withhold the cure from people who didn't share their data with us.

And finally, to a given person, it might be considered a security issue. People generally don't mind if someone has their data; they mind if the data can be used against them and/or linked to them personally.

Going back full circle to the technical issue, it's super hard to make data truly anonymous. For example, a recent Nature study, "Unique in the Crowd: the privacy bounds of human mobility" by Yves-Alexandre de Montjoye, et al., on a dataset of 1.5 million cell-phone users in Europe showed that just four points of reference were enough to individually identify 95 percent of the people.

Recently we've seen people up in arms about the way the NSA collects data about its citizens (not to mention non-US citizens). In fact, as this book was close to going to print, the Edward Snowden leak occurred. The response has been a very public and loud debate about the right to privacy with respect to our government. Considering how much information is bought and sold online about individuals through information warehousers and brokers like Acxiom—which leads not just to marketing but also insurance, job, and loan information—we might want to consider having that same conversation about our right to privacy with respect to private companies as well.

# MapReduce

Here we get insight into how David, as an engineer, thinks.

He revises the question we've asked before in this book: what *is* Big Data? It's a buzzword mostly, but it can be useful. He tried this as a working definition:

You're dealing with Big Data when you're working with data that doesn't fit into your computer unit. Note that makes it an evolving definition: Big Data has been around for a long time. The IRS had taxes before computers, so by our definition, the data they had didn't fit on their (nonexistent) computer unit.

Today, Big Data means working with data that doesn't fit in one computer. Even so, the size of Big Data changes rapidly. Computers have experienced exponential growth for the past 40 years. We have at least

10 years of exponential growth left (and people said the same thing 10 years ago).

Given this, is Big Data going to go away? Can we ignore it?

David claims we can't, because although the capacity of a given computer is growing exponentially, those same computers also *make* the data. The rate of new data is also growing exponentially. So there are actually two exponential curves, and they won't intersect any time soon.

Let's work through an example to show how hard this gets.

# Word Frequency Problem

Say you're told to find the most frequent words in the following list: red, green, bird, blue, green, red, red.

The easiest approach for this problem is inspection, of course. But now consider the problem for lists containing 10,000, or 100,000, or $10^9$ words.

The simplest approach is to list the words and then count their prevalence. Figure 14-1 shows an example code snippet from the language Go (*http://golang.org/*), which David loves and is helping to build (do you build a language?) and design at Google.

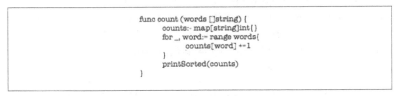

```
func count (words []string) {
    counts:- map[string]int[]
    for _, word:= range words{
        counts[word] +=1
    }
    printSorted(counts)
}
```

*Figure 14-1. Example code snippet from the language Go*

Because counting and sorting are fast, this scales to ~100 million words. The limit now is computer memory—if you think about it, you need to get all the words into memory twice: once when you load in the list of all words, and again when you build a way to associate a count for each word.

You can modify it slightly so it doesn't have to have all words loaded in memory—keep them on the disk and stream them in by using a channel instead of a list. A channel is something like a stream: you read in the first 100 items, then process them, then you read in the next 100 items.

But there's still a *potential* problem, because if every word is unique, and the list is super long, your program will still crash; it will still be too big for memory. On the other hand, this will *probably* work nearly all the time, because nearly all the time there *will* be repetition. Real programming is a messy game.

Hold up, computers nowadays are many-core machines; let's use them all! Then the bandwidth will be the problem, so let's compress the inputs, too. That helps, and moreover there are better alternatives that get complex. A heap of hashed values has a bounded size and can be well-behaved. A heap is something like a partially ordered set, and you can throw away super small elements to avoid holding everything in memory. This won't always work, but it will in most cases.

### Are You Keeping Up?
You don't need to understand all these details, but we want you to get a flavor for the motivation for why MapReduce is even necessary.

Now we can deal with on the order of 10 trillion words, using one computer.

Now say we have 10 computers. This will get us 100 trillion words. Each computer has 1/10th of the input. Let's get each computer to count up its share of the words. Then have each send its counts to one "controller" machine. The controller adds them up and finds the highest to solve the problem.

We can do this with hashed heaps, too, if we first learn network programming.

Now take a hundred computers. We can process *a thousand trillion words*. But then the "fan-in," where the results are sent to the controller, will break everything because of bandwidth problem. We need a tree, where every group of 10 machines sends data to one local controller, and then they all send back to super controller. This will probably work.

But… can we do this with 1,000 machines? No. It won't work. Because at that scale, one or more computer will fail. If we denote by $X$ the variable that exhibits whether a given computer is working, so $X = 0$ means it works and $X = 1$ means it's broken, then we can assume:

$$P(X = 0) = 1 - \epsilon$$

But this means, when we have 1,000 computers, the chance that no computer is broken is

$$(1 - \epsilon)^{1000}$$

which is generally pretty small even if $\epsilon$ is small. So if $\epsilon = 0.001$ for each individual computer, then the probability that all 1,000 computers work is 0.37, less than even odds. This isn't sufficiently robust.

What to do?

We address this problem by talking about fault tolerance for distributed work. This usually involves replicating the input (the default is to have three copies of everything), and making the different copies available to different machines, so if one blows, another one will still have the good data. We might also embed checksums in the data, so the data itself can be audited for errors, and we will automate monitoring by a controller machine (or maybe more than one?).

In general we need to develop a system that detects errors and restarts work automatically when it detects them. To add efficiency, when some machines finish, we should use the excess capacity to rerun work, again checking for errors.

 Q: Wait, I thought we were counting things?! This seems like some other awful rat's nest we've gotten ourselves into.

A: It's always like this. You cannot reason about the efficiency of fault tolerance easily; everything is complicated. And note, efficiency is just as important as correctness, because a thousand computers are worth more than your salary.

It's like this:

- The first 10 computers are easy;
- The first 100 computers are hard; and
- The first 1,000 computers are impossible.

There's really no hope.

Or at least there wasn't until about eight years ago. At Google now, David uses 10,000 computers regularly.

## Enter MapReduce

In 2004 Jeff and Sanjay published their paper "MapReduce: Simplified Data Processing on Large Clusters" (and here's another one on the underlying filesystem (*http://research.google.com/archive/gfs.html*)).

MapReduce allows us to stop thinking about fault tolerance; it is a platform that does the fault tolerance work for us. Programming 1,000 computers is now easier than programming 100. It's a library to do fancy things.

To use MapReduce, you write two functions: a mapper function, and then a reducer function. It takes these functions and runs them on many machines that are local to your stored data. All of the fault tolerance is automatically done for you once you've placed the algorithm into the map/reduce framework.

The mapper takes each data point and produces an ordered pair of the form (key, value). The framework then sorts the outputs via the "shuffle," and in particular finds all the keys that match and puts them together in a pile. Then it sends these piles to machines that process them using the reducer function. The reducer function's outputs are of the form (key, new value), where the new value is some aggregate function of the old values.

So how do we do it for our word counting algorithm? For each word, just send it to the ordered pair with the key as that word and the value being the integer 1. So:

```
red ---> ("red", 1)
blue ---> ("blue", 1)
red ---> ("red", 1)
```

Then they go into the "shuffle" (via the "fan-in") and we get a pile of ("red", 1)'s, which we can rewrite as ("red", 1, 1). This gets sent to the reducer function, which just adds up all the 1's. We end up with ("red", 2), ("blue", 1).

The key point is: one reducer handles *all the values for a fixed key*.

Got more data? Increase the number of map workers and reduce workers. In other words, do it on more computers. MapReduce flattens the complexity of working with many computers. It's elegant, and people use it even when they shouldn't (although, at Google it's not so crazy to assume your data could grow by a factor of 100 overnight). Like all tools, it gets overused.

Counting was one easy function, but now it's been split up into two functions. In general, converting an algorithm into a series of Map-Reduce steps is often unintuitive.

For the preceding word count, distribution needs to be uniform. If all your words are the same, they all go to one machine during the shuffle, which causes huge problems. Google has solved this using hash buckets heaps in the mappers in one MapReduce iteration. It's called CountSketch, and it is built to handle odd datasets.

At Google there's a real-time monitor for MapReduce jobs, a box with *shards* that correspond to pieces of work on a machine. It indicates through a bar chart how the various machines are doing. If all the mappers are running well, you'd see a straight line across. Usually, however, everything goes wrong in the reduce step due to nonuniformity of the data; e.g., lots of values on one key.

The data preparation and writing the output, which take place behind the scenes, take a long time, so it's good to try to do everything in one iteration. Note we're assuming a distributed filesystem is already there —indeed we have to use MapReduce to get data to the distributed filesystem—once we start using MapReduce, we can't stop.

Once you get into the optimization process, you find yourself tuning MapReduce jobs to shave off nanoseconds from repetitive processes because you're dealing with petabytes of data. These are order shifts worthy of physicists. This optimization is almost all done in C++. It's highly optimized code, and we try to scrape out every ounce of power we can.

# Other Examples of MapReduce

Counting words is the most basic example of MapReduce. Let's look at another to start getting more of a feel for it. The key attribute of a problem that can be solved with MapReduce is that the data can be distributed among many computers and the algorithm can treat each of those computers separately, i.e., one computer doesn't need to know what's going on with any other computer.

Here's another example where you could use MapReduce. Let's say you had tons of timestamped event data and logs of users' actions on a website. For each user, you might have {user_id, IP_address, zip code, ad_they_saw, did_they_click}. Suppose you wanted to count how many unique users saw ads from each zip code and how many clicked at least once.

How would you use MapReduce to handle this? You could run Map-Reduce keyed by zip code so that a record with a person living in zip code 90210 would get emitted to (90210,{1,1}) if that person saw an ad and clicked, or (90210,{0,1}) if they saw an ad and didn't click

What would this give you? At the reducer stage, this would count the total number of clicks and impressions by zip code producing output of the form (90210,{700,15530}), for example. But that's not what you asked. You wanted the number of unique users. This would actually require two MapReduces.

First use {zip_code,user} as the key and {clicks, impressions} as the value. Then, for example ({90210,user_5321},{0,1}) or ({90210,user_5321}, {1,1}). The reducer would emit a table that per user, per zip code, gave the counts of clicks and impressions. Your new records would now be {user, zipcode, number_clicks, number_impressions}.

Then to get the number of unique users from each zip code and how many clicked at least once, you'd need a second MapReduce job with zipcode as the key, and for each user emits {1, ifelse(clicks>0)} as the value.

So that was a second illustration of using MapReduce to count. But what about something more complicated like using MapReduce to implement a statistical model such as linear regression. Is that possible?

Turns out it is. Here's a 2006 paper that goes through how the Map-Reduce framework could be used to implement a variety of machine learning algorithms (*http://goo.gl/dkuSMt*). Algorithms that calculate sufficient statistics or gradients that depend upon calcuating expected values and summations can use the general approach described in this paper, because these calculations may be batched, and are expressible as a sum over data points.

## What Can't MapReduce Do?

Sometimes to understand what something *is*, it can help to understand what it *isn't*. So what *can't* MapReduce do? Well, we can think of lots of things, like give us a massage. But you'd be forgiven for thinking MapReduce can solve any data problem that comes your way.

But MapReduce isn't ideal for, say, iterative algorithms where you take a just-computed estimation and use it as input to the next stage in the computation—this is common in various machine learning algorithms that use steepest descent convergence methods. If you wanted to use MapReduce, it is of course possible, but it requires firing up many stages of the engine. Other newer approaches such as Spark might be better suited, which in this context means more efficient.

# Pregel

Just as a point of contrast, another algorithm for processing large-scale data was developed at Google called Pregel. This is a graph-based computational algorithm, where you can imagine the data itself has a graph-based or network-based structure. The computational algorithm allows nodes to communicate with other nodes that they are connected to. There are also aggregators that are nodes that have access to the information that all the nodes have, and can, for example, sum together or average any information that all the nodes send to them.

The basis of the algorithm is a series of supersteps that alternate between nodes sending information to their neighbors and nodes sending information to the aggregators. The original paper is online if you want to read more (*http://dl.acm.org/citation.cfm?id=1807184*). There's also an open source version of Pregel called Giraph (*http://giraph.apache.org/*).

# About Josh Wills

Josh Wills is Cloudera's director of data science, working with customers and engineers to develop Hadoop-based solutions across a wide range of industries. More on Cloudera and Hadoop to come. Prior to joining Cloudera, Josh worked at Google, where he worked on the ad auction system and then led the development of the analytics infrastructure used in Google+. He earned his bachelor's degree in mathematics from Duke University and his master's in operations research from The University of Texas at Austin.

Josh is also known for pithy data science quotes, such as: "I turn data into awesome" and the one we saw way back in the start of the book: "data scientist (noun): Person who is better at statistics than any software engineer and better at software engineering than any statistician." Also this gem: "I am Forrest Gump, I have a toothbrush, I have a lot of data and I scrub."

Josh primed his topic with a thought experiment first.

# Thought Experiment

How would you build a human-powered airplane? What would you do? How would you form a team?

Maybe you'd run an X prize (*http://www.xprize.org/*) competition. This is exactly what some people did, for $50,000, in 1950. It took 10 years for someone to win it. The story of the winner is useful because it illustrates that sometimes you are solving the wrong problem.

Namely, the first few teams spent years planning, and then their planes crashed within seconds. The winning team changed the question to: how do you build an airplane you can put back together in four hours after a crash? After quickly iterating through multiple prototypes, they solved this problem in six months.

# On Being a Data Scientist

Josh had some observations about the job of a data scientist. A data scientist spends all their time doing data cleaning and preparation— a full 90% of the work is this kind of data engineering. When deciding between solving problems and finding insights, a data scientist solves problems. A bit more on that: start with a problem, and make sure you have something to optimize against. Parallelize everything you do.

It's good to be smart, but being able to learn fast is even better: run experiments quickly to learn quickly.

## Data Abundance Versus Data Scarcity

Most people think in terms of scarcity. They are trying to be conservative, so they throw stuff away. Josh keeps everything. He's a fan of reproducible research, so he wants to be able to rerun any phase of his analysis. He keeps everything. This is great for two reasons. First, when he makes a mistake, he doesn't have to restart everything. Second, when he gets new sources of data, it's easy to integrate them in the point of the flow where it makes sense.

## Designing Models

Models always turn into crazy Rube Goldberg machines, a hodge-podge of different models. That's not necessarily a bad thing, because if they work, they work. Even if you start with a simple model, you eventually add a hack to compensate for something. This happens over and over again; it's the nature of designing the model.

### Mind the gap

The thing you're optimizing with your model isn't the same as the thing you're optimizing for your business.

Example: friend recommendations on Facebook don't optimize you accepting friends, but rather *maximizing the time you spend on Facebook*. Look closely: the suggestions are surprisingly highly populated by attractive people of the opposite sex.

How does this apply in other contexts? In medicine, they study the effectiveness of a drug instead of the health of the patients. They typically focus on success of surgery rather than well-being of the patient.

# Economic Interlude: Hadoop

Let's go back to MapReduce and Hadoop for a minute. When Josh graduated in 2001, there were two options for file storage—databases and filers—which Josh compared on four dimensions: schema, processing, reliability, and cost (shown in Table 14-1).

*Table 14-1. Options for file storage back in 2001*

|  | Databases | Filers |
| --- | --- | --- |
| Schema | Structured | No schemas |
| Processing | Intensive processing done where data is stored | No data processing capability |
| Reliability | Somewhat reliable | Reliable |
| Cost | Expensive at scale | Expensive at scale |

Since then we've started generating lots more data, mostly from the Web. It brings up the natural idea of a data economic indicator: *return on byte*. How much value can we extract from a byte of data? How much does it cost to store? If we take the ratio, we want it to be bigger than one, or else we discard the data.

Of course, this isn't the whole story. There's also a Big Data economic law, which states that *no individual record is particularly valuable, but having every record is incredibly valuable.* So, for example, for a web index, recommendation system, sensor data, or online advertising, one has an enormous advantage if one has all the existing data, even though each data point on its own isn't worth much.

## A Brief Introduction to Hadoop

Back before Google had money, it had crappy hardware. So to handle all this data, it came up with idea of copying data to multiple servers. It did this physically at the time, but then automated it. The formal automation of this process was the genesis of GFS.

There are two core components to Hadoop, which is the open source version of Google's GFS and MapReduce. (You can read more about the origin stories of Hadoop elsewhere. We'll give you a hint. A small open source project called Nutch and Yahoo! were involved.) The first component is the distributed filesystem (HDFS), which is based on the Google filesystem. The data is stored in large files, with block sizes of 64 MB to 256 MB. The blocks are replicated to multiple nodes in the cluster. The master node notices if a node dies. The second component is MapReduce, which David Crawshaw just told us all about.

Also, Hadoop is written in Java, whereas Google stuff is in C++. Writing MapReduce in the Java API is not pleasant. Sometimes you have to write lots and lots of MapReduces. However, if you use Hadoop streaming, you can write in Python, R, or other high-level languages. It's easy and convenient for parallelized jobs.

## Cloudera

Cloudera was cofounded by Doug Cutting, one of the creators of Hadoop, and Jeff Hammerbacher, who we mentioned back in Chapter 1 because he co-coined the job title "data scientist" when he worked at Facebook and built the data science team there.

Cloudera is like Red Hat for Hadoop, by which we mean they took an open source project and built a company around it. It's done under the aegis of the Apache Software Foundation. The code is available for free, but Cloudera packages it together, gives away various distributions for free, and waits for people to pay for support and to keep it up and running.

Apache Hive (*http://hive.apache.org/*) is a data warehousing system on top of Hadoop. It uses a SQL-based query language (includes some MapReduce-specific extensions), and it implements common join and aggregation patterns. This is nice for people who know databases well and are familiar with stuff like this.

# Back to Josh: Workflow

With respect to how Josh would approach building a pipeline using MapReduce, he thinks of his basic unit of analysis as a record. We've repeatedly mentioned "timestamped event data," so you could think of a single one of those as a record, or you could think of transaction records that we discussed in fraud detection or credit card transactions. A typical workflow would then be something like:

1. Use Hive (a SQL-like language that runs on Hadoop) to build records that contain everything you know about an entity (say a person) (intensive MapReduce stuff).

2. Write Python scripts to process the records over and over again (faster and iterative, also MapReduce).

3. Update the records when new data arrives.

Note the scripts in phase 2 are typically map-only jobs, which makes parallelization easy.

Josh prefers standard data formats: text is big and takes up space. Thrift, Avro, and protocol buffers are more compact, binary formats. He also encourages you to use the code and metadata repository Git-Hub (*https://github.com/*). He doesn't keep large data files in Git.

## So How to Get Started with Hadoop?

If you are working at a company that has a Hadoop cluster, it's likely that your first experience will be with Apache Hive, which provides a SQL-style abstraction on top of HDFS and MapReduce. Your first MapReduce job will probably involve analyzing logs of user behavior in order to get a better understanding of how customers are using your company's products.

If you are exploring Hadoop and MapReduce on your own for building analytical applications, there are a couple of good places to start. One option is to build a recommendation engine using Apache Mahout, a collection of machine learning libraries and command-line tools that works with Hadoop. Mahout has a collaborative filtering engine called Taste that can be used to create recommendations given a CSV file of user IDs, item IDs, and an optional weight that indicates the strength of the relationship between the user and the item. Taste uses the same recommendation algorithms that Netflix and Amazon use for building recommendations for their users.

CHAPTER 15

# The Students Speak

Every algorithm is editorial.

— Emily Bell (director of the Tow Center for Digital Journalism at
Columbia's Graduate School of Journalism)

We invited the students who took Introduction to Data Science version
1.0 to contribute a chapter to the book. They chose to use their chapter
to reflect on the course and describe how they experienced it. Contributors to this chapter are Alexandra Boghosian, Jed Dougherty,
Eurry Kim, Albert Lee, Adam Obeng, and Kaz Sakamoto.

## Process Thinking

When you're learning data science, you can't start anywhere except the
cutting edge.

An introductory physics class will typically cover mechanics, electric-
ity, and magnetism, and maybe move on to some more "modern" sub-
jects like special relativity, presented broadly in order of ascending
difficulty. But this presentation of accumulated and compounded
ideas in an aggregated progression doesn't give any insight into, say,
how Newton actually came up with a differential calculus. We are not
taught about his process; how he got there. We don't learn about his
tools or his thoughts. We don't learn which books he read or whether
he took notes. Did he try to reproduce other people's proofs? Did he
focus on problems that followed from previous writing? What exactly
made him think, "I just can't do this without infinitesimals?" Did
Newton need scratch paper? Or did the ideas emerge somehow fully
formed when he saw an apple drop? These things aren't taught, but

they have to be learned; this process is what the fledgling scientists will actually have to do.

Rachel started the first Introduction to Data Science class with a hefty caveat. Data science is still being defined, in both industry and academia. In each subsequent lecture, we learned about substantive problems and how people decide what problems to study. Substantively, the weekly lectures covered the tools and techniques of a data scientist, but each lecturer had their own style, background, and approach to each problem. In almost every class, the speaker would say something like, "I don't know what you guys have covered so far, but…" The lectures were discretized in this way, and it was our job to interpolate a continuous narrative about data science. We had to create our own meaning from the course, just as data scientists continue to construct the field to which they belong.

That is not to say that Rachel left us in the dark. On the first day of class, she proposed a working definition of data science. A data scientist, she said, was a person whose aptitude was distributed over the following: mathematics, statistics, computer science, machine learning, visualization, communication, and domain expertise. We would soon find out that this was only a prior in our unfolding Bayesian understanding. All the students and each lecturer evaluated themselves in terms of this definition, providing at once a diverse picture of the data science community and a reference point throughout the course. The lecturers came from academia, finance, established tech companies, and startups. They were graduate school dropouts, Kaggle competition winners, and digital artists. Each provided us with a further likelihood ratio. The class itself became sort of an iterative definition of data science.

But we didn't just listen to people talk about their jobs week after week. We learned the tools of the trade with lots of difficult, head-to-table homework. Sometimes the assignments required us to implement the techniques and concepts we had discussed in the lectures. Other times we had to discover and use skills we didn't even know existed.

What's more, we were expected to deal with messy real-world data. We often worked on industry-related problems, and our homework was to be completed in the form of a clear and thoughtful report—something we could pride ourselves in presenting to an industry professional. Most importantly we often had little choice but to reach out beyond our comfort zones to one another to complete assignments. The social nature of data science was emphasized, and in addition to the formal groupings for assignments and projects, Rachel would often take us to the bar across the street along with whoever had presented that evening.[1] We worked and drank together throughout the course, learning from one another and building our skills together.

# Naive No Longer

Our reckoning came on the third subsection of our second homework assignment, "Jake's Exercise: Naive Bayes for Article Classification" on page 108. It required us to download 2,000 articles from the *New York Times*—which only allowed a pull of 20 articles at a time—and train a simple Naive Bayes classifier to sort them by the section of the newspaper in which they appeared. Acquiring the articles was only half the challenge. While there are packages for many languages that implement Naive Bayes, we had to write our own code, with nothing more than a few equations for guidance. Not that using an existing package would not have helped us. Unlike them, our version had to include tunable regularization hyperparameters. It also demanded that we classify the articles across five categories instead of two. As we were told, simple Naive Bayes isn't particularly naive, nor very Bayesian. Turns out it's not that simple, either. Still, those of us who stayed in the class through the pain of spending 40 hours "polishing" 300 lines of disgustingly hacky R code got the pleasure of seeing our models graze 90% predictive accuracy. We were instantly hooked. Well, maybe it was the sunk cost. But still, we were hooked. Figure 15-1 shows one student's solution to the homework.

---

1. Rachel's note: it was a graduate-level course. I made sure students were of age and that there were non-alcoholic options.

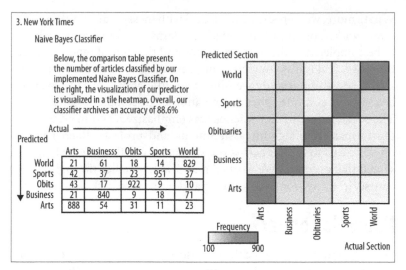

*Figure 15-1. Part of a student's solution to a homework assignment*

The Kaggle competition that made up half of our final project was a particular opportunity to go off the beaten track. The final was designed as a competition between the members of the class to create an essay-grading algorithm. The homework often simulated the data scientist's experience in industry, but Kaggle competitions could be described as the dick-measuring contests of data science. It encouraged all of the best data science practices we had learned while putting us in the thick of a quintessential data science activity. One of the present authors' solutions ended up including as features (among others) the number of misspellings, the number of words on the Dale-Chall list of words a fourth-grader should understand, TF-IDF vectors of the top 50 words, and the fourth root of the number of words in the essay. Don't ask why. The model used both random forest and gradient boosted models, and implemented stochastic hyperparameter optimization, training fifty thousand models across thousands of hours on Amazon EC2 instances. It worked OK.

# Helping Hands

In one of the first classes, we met guest lecturer Jake Hofman. Do you remember seeing your first magic trick? Jake's lecture? Yeah, it was like that—as prestidigitatory as any close-up act. By using a combination of simple Unix utilities and basic data, he built a Naive Bayes spam classifier before our very eyes. After writing a few math equations on the board, he demonstrated some of his "command-line-fu" to parse the publicly available Enron emails he downloaded live.

In the weekly lectures, we were presented with standing-ovation-inducing performers. But it was with the help of Jared Lander and Ben Reddy, who led our supplementary lab sessions, that we stayed afloat in the fast-paced class. They presented to us the mechanics of data science. We covered the gamut: statistical concepts ranging from linear regression to the mechanics behind the random forest algorithm. And many of us were introduced to new tools: regular expressions, LaTeX, SQL, R, the shell, Git, and Python. We were given the keys to new sources of data through APIs and web-scraping.

Generally, the computer scientists in the class had to quickly pick up the basics of theory-based feature selection and using R. Meanwhile, the social scientists had to understand the mechanics of a database and the differences between globally and locally scoped variables, and the finance majors had to learn ethics. We all had our burdens. But as we all methodically yet mistakenly constructed for loops in R and considered the striking inefficiencies of our code, our bags of tricks became a little heavier. Figure 15-2 shows one of the lessons or pitfalls of this method:

```
44   #WARNING THIS TAKES A LONG TIME AND MAKES YOUR COMPUTER GET REALLY HOT
45   detailed_results <- predict(model,as.matrix(testmatrix),type="raw");
```

*Figure 15-2. Lesson learned*

And as our skills improved, we were increasingly able to focus on the analysis of the data. Eventually, we didn't even see the code—just the ideas behind it.

But how could we figure these things out on our own? Could any one of us do it all?

It took hours of frustrating errors and climbing learning curves to appreciate the possibilities of data science. But to actually finish our homework on time, we had to be willing to learn from the different backgrounds of the students in the class.

In fact, it became crucial for us to find fellow students with complementary skills to complete our assignments. Rachel forced group work upon us, not by requiring it directly, but by assigning massive yet discretizable assignments. It turned out that she meant for us to know that practicing data science is inherently a collective endeavor. In the beginning of the course, Rachel showed us a hub-and-spoke network diagram. She had brought us all together and so was at the center. The spokes connected each of us to her. It became her hope that new friendships/ideas/projects/connections would form during the course.

It's perhaps more important in an emergent field than in any other to be part of a community. For data science in particular, it's not just useful to your career—it's essential to your practice. If you don't read the blogs, or follow people on Twitter, or attend meetups, how can you find out about the latest distributed computing software, or a refutation of the statistical approach of a prominent article? The community is so tight-knit that when Cathy was speaking about MapReduce at a meetup in April, she was able to refer a question to an audience member, Nick Avteniev—easy and immediate references to the experts of the field is the norm. Data science's body of knowledge is changing and distributed, to the extent that the only way of finding out what you should know is by looking at what other people know. Having a bunch of different lecturers kickstarted this process for us. All of them answered our questions. All gave us their email addresses. Some even gave us jobs.

Having listened to and conversed with these experts, we formed more questions. How can we create a time series object in R? Why do we keep getting errors in our plotting of our confusion matrix? What the heck is a random forest? Here, we not only looked to our fellow students for answers, but we went to online social communities such as Stack Overflow, Google Groups, and R bloggers. It turns out that there is a rich support community out there for budding data scientists like us trying to make our code run. And we weren't just getting answers from others who had run into the same problems before us. No, these questions were being answered by the pioneers of the methods. People like Hadley Wickham, Wes McKinney, and Mike Bostock were providing support for the packages they themselves wrote. Amazing.

# Your Mileage May Vary

It's not as if there's some platonic repository of perfect data science knowledge that you can absorb by osmosis. There are various good practices from various disciplines, and different vocabularies and interpretations for the same method (is the regularization parameter a prior, or just there for smoothing? Should we choose it on principled grounds or in order to maximize fit?) There is no institutionalized knowledge because there are no institutions, and that's why the structure of interactions matters: you can can create your own institutions. You choose who you are influenced by, as Gabriel Tarde put it (via Bruno Latour, via Mark Hansen):

> When a young farmer, facing the sunset, does not know if he should believe his school master asserting that the fall of the day is due to the movement of the earth and not of the sun, or if he should accept as witness his senses that tell him the opposite, in this case, there is one imitative ray, which, through his school master, ties him to Galileo.
>
> — Gabriel Tarde

Standing on the shoulders of giants is all well and good, but before jumping on someone's back you might want to make sure that they can take the weight. There is a focus in the business world to use data science to sell advertisements. You may have access to the best dataset in the world, but if the people employing you only want you to find out how to best sell shoes with it, is it really worth it?

As we worked on assignments and compared solutions, it became clear that the results of our analyses could vary widely based on just a few decisions. Even if you've learned all the steps in the process from hypothesis-building to results, there are so many ways to do each step that the number of possible combinations is huge. Even then, it's not as simple as piping the output of one command into the next. Algorithms are editorial, and the decision of which algorithm and variables to use is even more so.

Claudia Perlich from Media 6 Degrees (M6D) was a winner of the prestigious KDD Cup in 2003, 2007, 2008, 2009, and now can be seen on the coordinating side of these competitions. She was generous enough to share with us the ins and outs of the data science circuit and the different approaches that you can take when making these editorial decisions. In one competition to predict hospital treatment outcomes, she had noticed that patient identifiers had been assigned sequentially, such that all the patients from particular clinics had sequential numbers. Because different clinics treated patients with different severities of condition, the patient ID turned out to be a great predictor for the outcome in question. Obviously, the inclusion of this data leakage was unintentional. It made the competition trivial. But in the real world, perhaps it should actually be used in models; after all, the clinic that doctors and patients choose should probably be used to predict their outcomes.

David Madigan emphasized the ethical challenges of editorial decisions in this emerging domain by showing us how observational studies in the pharmaceutical industry often yield vastly different results. (Another example is the aspirin plot he showed.) He emphasized the importance of not removing oneself from the real data. It is not enough to merely tweak models and methods and apply them to datasets.

The academic world has a bit of the same problem as the business world, but for different reasons. The different bits of data science are so split between disciplines that by studying them individually it becomes nearly impossible to get a holistic view of how these chunks fit together, or even that they could fit together. A purely academic approach to data science can sterilize and quantize it to the point where you end up with the following, which is an actual homework problem from the chapter "Linear Methods for Regression" in *The Elements of Statistical Learning*:

---

Ex. 3.2 Given data on two variables $X$ and $Y$, consider fitting a cubic polynomial regression model $f(X) = \sum_{j=0}^{3} \beta_j X^j$. In addition to plotting the fitted curve, you would like a 95% confidence band about the curve. Consider the following two approaches:

1. At each point $x_0$, form a 95% confidence interval for the linear function $\alpha^T \beta = \sum_{j=0}^{3} \beta_j x_0^j$.

2. Form a 95% confidence set for $\beta$, which in turn generates confidence intervals for $f(x_0)$.

How do these approaches differ? Which band is likely to be wider? Conduct a small simulation experiment to compare the two methods.

---

This is the kind of problem you might be assigned in a more general machine learning or data mining class. As fledgling data scientists, our first reaction is now skeptical. At which point in the process of doing data science would a problem like this even present itself? How much would we have to have already done to get to this point? Why are we considering these two variables in particular? How come we're given data? Who gave it to us? Who's paying them? Why are we calculating 95% confidence intervals? Would another performance metric be better? Actually, who cares how well we perform on what is, essentially, our training data?

This is not completely fair to Hastie and his coauthors. They would probably argue that if students wanted to learn about data scraping and organization, they should get a different book that covers those topics—the difference in the problems shows the stark contrast in approach that this class took from normal academic introductory courses. The philosophy that was repeatedly pushed on us was that understanding the statistical tools of data science without the context of the larger decisions and processes surrounding them strips them of much of their meaning. Also, you can't just be told that real data is messy and a pain to deal with, or that in the real world no one is going to tell you exactly which regression model to use. These issues—and the intuition gained from working through them—can only be understood through experience.

# Bridging Tunnels

As fledgling data scientists, we're not—with all due respect to Michael Driscoll—exactly civil engineers. We don't necessarily have a grand vision for what we're doing; there aren't always blueprints. Data scientists are adventurers, we know what we're questing for, we've some tools in the toolkit, and maybe a map, and a couple of friends. When we get to the castle, our princess might be elsewhere, but what matters is that along the way we stomped a bunch of Goombas and ate a bunch of mushrooms, and we're still spitting hot fire. If science is a series of pipes, we're not plumbers. We're the freaking Mario Brothers.

# Some of Our Work

The students improved on the original data science profile from back in Chapter 1 in Figure 15-3 and created an infographic for the growing popularity of data science in universities in Figure 15-4, based on information available to them at the end of 2012.

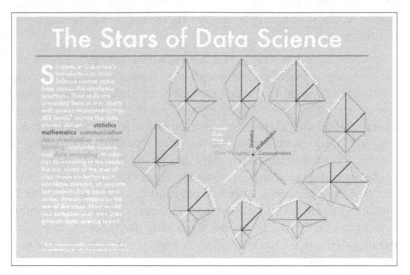

Figure 15-3. The Stars of Data Science (a collaborative effort among many students including Adam Obeng, Eurry Kim, Christina Gutierrez, Kaz Sakamoto, and Vaibhav Bhandari)

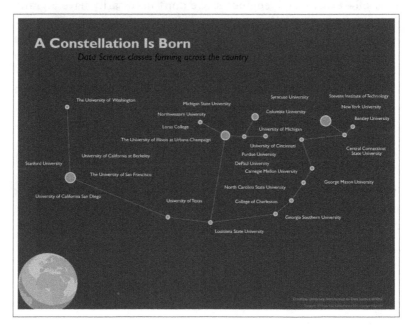

Figure 15-4. A Constellation is Born (Kaz Sakamoto, Eurry Kim and Vaibhav Bhandari created this as part of a larger class collaboration)

# CHAPTER 16

# Next-Generation Data Scientists, Hubris, and Ethics

We want to wrap up by thinking about what just happened and what we hope for you going forward.

## What Just Happened?

The two main goals of this book were to communicate what it's like to be a data scientist and to teach you how to do some of what a data scientist does.

We'd like to think we accomplished both of these goals.

The various contributors for the chapters brought multiple first-hand accounts of what it's like to be a data scientist, which addressed the first goal. As for the second goal, we are proud of the breadth of topics we've covered, even if we haven't been able to be as thorough as we'd like to be.

It's possible one could do better than we have, so the next sidebar gives you something to think about.

# What Is Data Science (Again)?

We revisited this question again and again throughout the book. It was the theme of the book, the central question, and the mantra.

Data science could be defined simply as what data scientists do, as we did earlier when we talked about profiles of data scientists. In fact, before Rachel taught the data science course at Columbia, she wrote up a list of all the things data scientists do and didn't want to show it to anyone because it was overwhelming and disorganized. That list became the raw material of the profiles. But after speaking to different people after the course, she's found that they actually like looking at this list, so here it is:

- Exploratory data analysis
- Visualization (for exploratory data analysis and reporting)
- Dashboards and metrics
- Find business insights
- Data-driven decision making
- Data engineering/Big Data (MapReduce, Hadoop, Hive, and Pig)
- Get the data themselves
- Build data pipelines (logs→mapreduce→dataset→join with other data→mapreduce→scrape some data→join)
- Build products instead of describing existing product usage
- Hack
- Patent writing
- Detective work
- Predict future behavior or performance
- Write up findings in reports, presentations, and journals
- Programming (proficiency in R, Python, C, Java, etc.)
- Conditional probability
- Optimization
- Algorithms, statistical models, and machine learning
- Tell and interpret stories
- Ask good questions
- Investigation
- Research
- Make inferences from data
- Build data products
- Find ways to do data processing, munging, and analysis at scale
- Sanity checking
- Data intuition
- Interact with domain experts (or be a domain expert)
- Design and analyze experiments
- Find correlation in data, and try to establish causality

But at this point, we'd like to go a bit further than that, to strive for something a bit more profound.

Let's define data science beyond a set of best practices used in tech companies. That's the definition we started with at the start of the book. But after going through this exploration, now consider data science to be beyond tech companies to include all other domains: neuroscience, health analytics, eDiscovery, computational social sciences, digital humanities, genomics, policy...to encompass the space of all problems that could possibly be solved with data *using* a set of best practices discussed in this book, some of which were initially established in tech companies. Data science happens both in industry and in academia, i.e., *where* or *what domain* data science happens in is not the issue—rather, defining it as a "problem space" with a corresponding "solution space" in algorithms and code and data is the key.

In that vein, start with this: data science is a set of best practices used in tech companies, working within a broad space of problems that could be solved with data, possibly even at times deserving the name science. Even so, it's sometimes nothing more than pure hype, which we need to guard against and avoid adding to.

# What Are Next-Gen Data Scientists?

> The best minds of my generation are thinking about how to make people click ads... That sucks.
>
> — Jeff Hammerbacher

Ideally the generation of data scientists-in-training are seeking to do more than become technically proficient and land a comfy salary in a nice city—although those things would be nice. We'd like to encourage the next-gen data scientists to become problem solvers and question askers, to think deeply about appropriate design and process, and to use data responsibly and make the world better, not worse. Let's explore those concepts in more detail in the next sections.

## Being Problem Solvers

First, let's discuss the technical skills. Next-gen data scientists should strive to have a variety of hard skills including coding, statistics, machine learning, visualization, communication, and math. Also, a solid foundation in writing code, and coding practices such as paired programming, code reviews, debugging, and version control are incredibly valuable.

It's never too late to emphasize exploratory data analysis as we described in Chapter 2 and conduct feature selection as Will Cukierski emphasized. Brian Dalessandro emphasized the infinite models a data scientist has to choose from—constructed by making choices about which classifier, features, loss function, optimization method, and evaluation metric to use. Huffaker discussed the construction of features or metrics: transforming the variables with logs, constructing binary variables (e.g., the user did this action five times), and aggregating and counting. As a result of perceived triviality, all this stuff is often overlooked, when it's a critical part of data science. It's what Dalessandro called the "Art of Data Science."

Another caution: many people go straight from a dataset to applying a fancy algorithm. But there's a huge space of important stuff in between. It's easy to run a piece of code that predicts or classifies, and to declare victory when the algorithm converges. That's not the hard part. The hard part is doing it well and making sure the results are correct and interpretable.

### What Would a Next-Gen Data Scientist Do?

Next-gen data scientists don't try to impress with complicated algorithms and models that don't work. They spend a lot more time trying to get data into shape than anyone cares to admit —maybe up to 90% of their time. Finally, they don't find religion in tools, methods, or academic departments. They are versatile and interdisciplinary.

## Cultivating Soft Skills

Tons of people can implement k-nearest neighbors, and many do it badly. In fact, almost everyone starts out doing it badly. What matters isn't where you start out, it's where you go from there. It's important that one cultivates good habits and that one remains open to continuous learning.

Some habits of mind that we believe might help solve problems[1] are persistence, thinking about thinking, thinking flexibly, striving for accuracy, and listening with empathy.

Let's frame this somewhat differently: in education in traditional settings, we focus on answers. But what we probably *should* focus on, or at least emphasize more strongly, is how students behave *when they don't know the answer*. We need to have qualities that help us find the answer.

Speaking of this issue, have you ever wondered why people *don't* say "I don't know" when they don't know something? This is partly explained through an unconscious bias called the Dunning-Kruger effect.

Basically, people who are bad at something have no idea that they are bad at it and overestimate their confidence. People who are super good at something underestimate their mastery of it. Actual competence may weaken self-confidence. Keep this in mind and try not to over- or underestimate your abilities—give yourself reality checks by making sure you can code what you speak and by chatting with other data scientists about approaches.

---

### Thought Experiment Revisited: Teaching Data Science

How would you design a data science class around *habits of mind* rather than technical skills? How would you quantify it? How would you evaluate it? What would students be able to write on their resumes?

---

## Being Question Askers

People tend to overfit their models. It's human nature to want your baby to be awesome, and you could be working on it for months, so yes, your feelings can become pretty maternal (or paternal).

It's also human nature to underestimate the bad news and blame other people for bad news, because from the parent's perspective, nothing one's own baby has done or is capable of is bad, unless someone else

---

1. Taken from the book *Learning and Leading with Habits of Mind*, edited by Arthur L. Costa and Bena Kallick (ACSD).

somehow made them do it. How do we work against this human tendency?

Ideally, we'd like data scientists to merit the word "scientist," so they act as someone who tests hypotheses and welcomes challenges and alternative theories. That means: shooting holes in our own ideas, accepting challenges, and devising tests as scientists rather than defending our models using rhetoric or politics. If someone thinks they can do better, then let them try, and agree on an evaluation method beforehand. Try to make things objective.

Get used to going through a standard list of critical steps: Does it have to be this way? How can I measure this? What is the appropriate algorithm and why? How will I evaluate this? Do I really have the skills to do this? If not, how can I learn them? Who can I work with? Who can I ask? And possibly the most important: how will it impact the real world?

Second, get used to asking other people questions. When you approach a problem or a person posing a question, start with the assumption that you're smart, and don't assume the person you're talking to knows more or less than you do. You're not trying to prove anything —you're trying to find out the truth. Be curious like a child, not worried about appearing stupid. Ask for clarification around notation, terminology, or process: Where did this data come from? How will it be used? Why is this the right data to use? What data are we ignoring, and does it have more features? Who is going to do what? How will we work together?

Finally, there's a really important issue to keep in mind, namely the classical statistical concept of causation versus correlation. Don't make the mistake of confusing the two. Which is to say, err on the side of assuming that you're looking at correlation.

### What Would a Next-Gen Data Scientist Do?

Next-gen data scientists remain skeptical—about models themselves, how they can fail, and the way they're used or can be misused. Next gen data scientists understand the implications and consequences of the models they're building. They think about the feedback loops and potential gaming of their models.

# Being an Ethical Data Scientist

You all are not just nerds sitting in the corner. You have increasingly important ethical questions to consider while you work.

We now have tons of data on market and human behavior. As data scientists, we bring not just a set of machine learning tools, but also our humanity, to interpret and find meaning in data and make ethical, data-driven decisions.

Keep in mind that the data generated by user behavior becomes the building blocks of data products, which simultaneously are used *by* users and *influence* user behavior. We see this in recommendation systems, ranking algorithms, friend suggestions, etc., and we will see it increasingly across sectors including education, finance, retail, and health. Things can go wrong with such feedback loops: keep the financial meltdown in mind as a cautionary example.

Much is made about predicting the future (see Nate Silver), predicting the present (see Hal Varian), and exploring causal relationships from observed data (the past; see Sinan Aral).

The next logical concept then is: models and algorithms are not only capable of predicting the future, but also of causing the future. That's what we can look forward to, in the best of cases, and what we should fear in the worst.

As an introduction to how to approach these issues ethically, let's start with Emanuel Derman's Hippocratic Oath of Modeling, which was made for financial modeling but fits perfectly into this framework:

- I will remember that I didn't make the world and that it doesn't satisfy my equations.
- Though I will use models boldly to estimate value, I will not be overly impressed by mathematics.
- I will never sacrifice reality for elegance without explaining why I have done so. Nor will I give the people who use my model false comfort about its accuracy. Instead, I will make explicit its assumptions and oversights.
- I understand that my work may have enormous effects on society and the economy, many of them beyond my comprehension.

Something that this oath does not take into consideration, but which you might have to as a data scientist, is the politics of working in industry. Even if you are honestly skeptical of your model, there's always the chance that it will be used the wrong way in spite of your warnings. So the Hippocratic Oath of Modeling is, unfortunately, insufficient in reality (but it's a good start!).

### What Would a Next-Gen Data Scientist Do?

Next-gen data scientists don't let money blind them to the point that their models are used for unethical purposes. They seek out opportunities to solve problems of social value and they try to consider the consequences of their models.

Finally, there are ways to do good: volunteer to work on a long-term project (more than a hackathon weekend) with DataKind (*http://data kind.org*).

There are also ways to be transparent: Victoria Stodden is working on RunMyCode (*http://www.runmycode.org/CompanionSite/*), which is all about making research open source and replicable.

We want step aside for a moment and let someone else highlight how important we think ethics—and vanquishing hubris—are to data science. Professor Matthew Jones, from Columbia's History Department, attended the course. He is an expert in the history of science, and he wrote up some of his thoughts based on the course. We've included them here as some very chewy food for thought.

## Data & Hubris

In the wake of the 2012 presidential election, data people, those they love, and especially those who idealize them, exploded in schaden-freude about the many errors of the traditional punditocracy. Computational statistics and data analysis had vanquished prognos-tication based on older forms of intuition, gut instinct, long-term journalistic experience, and the decadent web of Washington insiders. The apparent (*http://ti.me/GzJlhH*) success (*http://lat.ms/1hkOxki*) of the Obama team and others using quantitative prediction (*http://ti.me/GzJlhH*) revealed that a new age in political analysis has been cemented. Older forms of "expertise," now with scare quotes, were invited to take a long overdue retirement and to permit a new data-driven political analysis to emerge.

It's a compelling tale, with an easy and attractive bifurcation of old and new forms of knowledge. Yet good data scientists have been far more reflective about the dangers of throwing away existing domain knowledge and its experts entirely.

Origin stories add legitimacy to hierarchies of expertise. Data mining has long had a popular, albeit somewhat apocryphal, origin story: the surprising discovery, using an association algorithm, that men who buy diapers tend often to buy beer at the same time in drug stores. Traditional marketing people, with their quant folk psychologies and intuitions about business, were heretofore to be vanquished before what the press probably still called an "electronic brain." The story follows a classic template. Probability and statistics from their origins (*http://press.princeton.edu/titles/4295.html*) in the European Enlightenment have long challenged traditional forms of expertise: the pricing of insurance and annuities using data rather than reflection of character of the applicant entailed the diminution and disappearance of older experts. In the book (*http://goo.gl/v5JhxG*) that introduced the much beloved (or dreaded) epsilons and deltas into real analysis, the great mathematician Augustin-Louis Cauchy blamed statisticians for the French Revolution: "Let us cultivate the mathematical sciences with ardor, without wanting to extend them beyond their domain; and let us not imagine that one can attack history with formulas, nor give sanction to morality through theories of algebra or the integral calculus."

These narratives fit nicely into the celebration of disruption so central to Silicon Valley libertarianism, Schumpeterian capitalism, and certain variants of tech journalism. However powerful in extirpating rent-seeking forms of political analysis and other disciplines, the dichotomy mistakes utterly the real skills and knowledge that appear often to give the data sciences the traction they have. The preceding chapters—dedicated to the means for cultivating the diverse capacities of the data scientist—make mincemeat of any facile dichotomy of the data expert and the traditional expert. *Doing data science* has put a tempering of hubris, especially algorithmic hubris, at the center of its technical training.

Obama's data team explained (*http://goo.gl/sB8pGH*) that much of their success came from taking the dangers of hubris rather seriously, indeed, in building a technical system premised on avoiding the dangers of overestimation, from the choice and turning of algorithms to the redundancy of the backend and network systems: "I think the Republicans f**ked up in the hubris department," Harper Reed explained to Atlantic writer Alexis Madrigal. "I know we had the best technology team I've ever worked with, but we didn't know if it would work. I was incredibly confident it would work. I was betting a lot on it. We had time. We had resources. We had done what we thought would work, and it still could have broken. Something could have happened."

Debate over the value of "domain knowledge" has long polarized the data community. Much of the promise of unsupervised learning, after all, is overcoming a crippling dependence on our wonted categories of social and scientific analysis, as seen in one of many celebrations (*http://lat.ms/1hkOxki*) of the Obama analytics team. Daniel Wagner, the 29-year-old chief analytics officer, said:

> The notion of a campaign looking for groups such as "soccer moms" or "waitress moms" to convert is outdated. Campaigns can now pinpoint individual swing voters. White suburban women? They're not all the same. The Latino community is very diverse with very different interests. What the data permits you to do is to figure out that diversity.

In productive tension with this escape from deadening classifications, however, the movement to revalidate domain expertise within statistics seems about as old as formalized data mining.

In a now infamous *Wall Street Journal* article (*http://on.wsj.com/15LnZno*), Peggy Noonan mocked the job ad for the Obama analytics department: "It read like politics as done by Martians." The campaign was simply insufficiently human, with its war room both "high-tech and bloodless." Unmentioned went that the contemporaneous Romney ads read similarly.

Data science rests on algorithms but does not reduce to those algorithms. The use of those algorithms rests fundamentally on what sociologists of science call "tacit knowledge" (*http://amzn.to/19huR1W*)—practical knowledge not easily reducible to articulated rules, or perhaps impossible to reduce to rules at all. Using algorithms well is fundamentally a very human endeavor—something not particularly algorithmic.

No warning to young data padawans is as central as the many dangers of overfitting, the taking of noise for signal in a given training set; or, alternatively, learning too much from a training set to generalize properly. Avoiding overfitting requires a reflective use of algorithms. Algorithms are enabling tools requiring us to reflect more, not less. In 1997 Peter Huber explained, "The problem, as I see it, is not one of replacing human ingenuity by machine intelligence, but one of assisting human ingenuity by all conceivable tools of computer science and artificial intelligence, in particular aiding with the improvisation of search tools and with keeping track of the progress of an analysis." The word 'improvisation' is just right in pointing to mastery of tools, contextual reasoning, and the virtue of avoiding rote activity. The hubris one might have when using an algorithm must be tempered through a profound familiarity with that algorithm and its particular instantiation.

Reflection upon the splendors and miseries of existing models figured prominently in the Obama campaign's job ad (*http://goo.gl/3KuIuj*) mocked by Noonan:

- Develop and build statistical/predictive/machine learning models to assist in field, digital media, paid media and fundraising operations

- Assess the performance of previous models and determine when these models should be updated

- Design and execute experiments to test the applicability and validity of these models in the field [...]

The facile, automatic application of models is simply not at issue here: criticism and evaluation are. No Martian unfamiliar with territory could do this: existing data—of all kind—is simply too vital to pass up.

How to learn to improvise? In other words, what model would be best for educating the data scientist? Becoming capable of reflective improvisation with algorithms and large data demands the valorization of the muddling through, the wrangling, the scraping, the munging of poorly organized, incomplete, likely inconclusive data. The best fitting model for the training required is not narrow vocational education, but—of all things—the liberal arts in their original meaning.

For centuries, an art, such as mathematics or music, was called "liberal" just because it wasn't automatic, mechanical, purely repetitive, habitual. A liberal arts education is one for free people, in the sense of people mentally free to reflect upon their tools, actions, and customs, people not controlled by those tools, people free, then, to use or not to use their tools. The point holds for algorithms as much as literature—regurgitation need not apply in the creation of a data scientist worth the name. Neither should any determinism about technology. No data scientist need ever confuse the possibility of using a technology with the necessity of using it. In the sparse space of the infinite things one might do with data, only a few dense (and interesting) ethical pockets deserve our energy.

— Matthew Jones

# Career Advice

We're not short on advice for aspiring next-gen data scientists, especially if you've gotten all the way to this part of the book.

After all, lots of people ask us whether they should become data scientists, so we're pretty used to it. We often start out the advice session with two questions of our own.

1. **What are you optimizing for?**

   To answer that question, you need to know what you value. For example, you probably value money, because you need some minimum to live at the standard of living you want to, and you might even want a lot. This definitely rules out working on lots of cool projects that would be cool to have in the world but which nobody wants to pay for (but don't forget to look for grants for projects like those!). Maybe you value time with loved ones and friends— in that case you will want to rule out working at a startup where everyone works twelve hours a day and sleeps under their desks. Yes, places like that still totally exist.

   You might care about some combination of doing good in the world, personal fulfillment, and intellectual fulfillment. Be sure to weigh your options with respect to these individually. They're definitely not the same things.

   What are your goals? What do you want to achieve? Are you interested in becoming famous, respected, or somehow specifically acknowledged? Probably your personal sweet spot is some weighted function of all of the above. Do you have any idea what the weights are?

2. **What constraints are you under?**

   There might be external factors, outside of your control, like you might need to live in certain areas with your family. Consider also money and time constraints, whether you need to think about vacation or maternity/paternity leave policies. Also, how easy would it be to sell yourself? Don't be painted into a corner, but consider how to promote the positive aspects of yourself: your education, your strengths and weaknesses, and the things you can or cannot change about yourself.

There are many possible solutions that optimize what you value and take into account the constraints you're under. From our perspective, it's more about personal fit than what's the "best job" on the market. Different people want and need different things from their careers.

On the one hand, remember that whatever you decide to do is not permanent, so don't feel too anxious about it. You can always do something else later—people change jobs all the time. On the other hand, life is short, so always try to be moving in the right direction— optimize for what you care about and don't get stagnant.

Finally, if you feel your way of thinking or perspective is somehow different than what those around you are thinking, then embrace and explore that; you might be onto something.

# Index

*We'd like to hear your suggestions for improving our indexes. Send email to index@oreilly.com.*

APIs, warnings about, 107
area under the cumulative lift curve,
    126
area under the curve (AUC), 309
    goodness of, 301
Artificial Artificial Intelligence, 169
Artificial Intelligence (AI), 52
ASA, 8
association algorithms, 358
associations, 60
assumptions, 22
    explicit, 53
attributes, predicting, 204
autocorrelation, 159–162
automated statistician thought experi-
    ment, 92
averages, 280
Avro, 338

# B

backward elimination, 182
bagging, 190
base, 207
base rates, 121
Basic Machine Learning Algorithm
    Exercise, 86–90
Bayesian Information Criterion (BIC),
    183, 205
Beautiful Soup, 107
Before Us is the Salesman's House
    (Thorp/Hansen), 232–234
Bell Labs, 35
Bernoulli model, 109
Bernoulli network, 266
betweenness, 258
bias-variance, 192
biases, 20, 21
Big Data, 2, 4–6, 24
    assumptions and, 24–26
    defined, 24
    economic law, 336
    Facebook and, 2
    Google and, 2
    objectivity of, 25
binary classes, 80
Binder, Phillipe M., 45
bipartite graphs, 136, 257

black box algorithms, 53
bootstrap aggregating, 190
bootstrap samples, 190
Bruner, Jon, 269–271

# C

caret packages, 245
case attribute vs. social network data,
    254
causal effect, 287
causal graphs, 286
causal inference, 286
causal models, 146–147, 146
causal questions, 274
causality, 273–290
    A/B testing for evaluation, 280–
        283
    causal questions, 274
    clinical trials to determine, 279–
        280
    correlation vs., 274–276
    observational studies and, 283–
        289
    OK Cupid example, 276–279
    unit level, 285
    visualizing, 286–287
centrality measures, 257–259
    eigenvalue centrality, 264
    usefulness of, 258
channels, 327
    problems with, 328
chaos, 45
chaos simulation thought experiment,
    44
charity example, 321
checksums, 329
choosing classifiers, 115
classes, 119
classification, 54
    binary, 80
classifiers, 115–118
    choosing, 115
    Naive Bayes, 341
    spam, 119
clear rates, 249
Cleveland, William, 9, 35
click models, 118

clinical trials to determine causality, 279–280
closeness, 75, 143, 257
Cloudera, 334, 337
clustering algorithms, 13, 83
clustering data, 82
code samples
    for EDA, 39–40
    RealDirect exercise, 49
coding, 40
cold stat problems, 242
computational complexity, 203
conditional distribution, 32
conditional entropy, 187
confidence intervals, 53
confounders, 275, 278
    correcting for, in practice, 295
    stratification and, 294–296
confusion matrix, 97
connected graphs, 257
constraints, on algorithms
    at runtime, 116
    interpretability of, 117
    scalability, 117
    thought experiments for, 113–128
    understanding, 117
constructing features, 309
continuous density function, 31
continuous variables in decision trees, 188
converged algorithms, 212
Conway, Drew, 7, 9
correlated variables, 176
correlation vs. causality, 274–276
cosine similarity, 76, 202
Cosma Shalizi, 8
CountSketch, 331
Crawford, Kate, 21
Crawshaw, David, 324
creating competitions, 166
credit score example, 73–75
Cronkite Plaza (Thorp/Rubin/Hansen), 231
cross-validation, 67, 213
crowdsourcing
    DARPA and, 167
    distributive, 167
    InnoCentive and, 167

issues with, 167
Kaggle and, 167
Mechanical Turks vs., 169
organization, 168
Wikipedia and, 167
Cukier, Kenneth Neil, 5
Cukierski, William, 165
curse of dimensionality, 96, 202
curves, goodness of, 301

# D

Dalessandro, Brian, 113
DARPA, 167
data
    abundance vs. scarcity, 335
    clustering, 82
    extracting meaning from, 165–198
    financial, preparing, 148–149
    from the Web, 106–108
    geo-based location, 23
    grouping, 82
    growing influence of, 4
    images, 23
    in-sample, 146–147
    labels, defining, 242
    network, 23
    normalizing, 148
    objectivity of, 25
    out-of-sample, 146–147
    products, 5
    real-time streaming, 23
    records, 23
    sensor, 23
    sparsity issues, 242
    stratifying, 82
    text, 23
    traditional, 23
    transforming, 148
    types, 23
    web APIs for, 106–108
data corpus, 21
data engineering, 323–338
    data abundance vs. scarcity, 335
    Hadoop and, 335–337
    MapReduce, 326
    models, designing, 335
    Pregel, 333

out-of-sample data, 146–147
S&P index example, 151
volatility measurements in, 153–155
finite differences, 183
Firebug extension, 107
Firefox, 107
Flickr, 108
FOIA requests, 269
Foreign Affairs (magazine), 5
forward selection, 182
Fourier coefficients, 247
fraud, data visualization and, 237–240
    detecting with machine learning, 237–240
    performance estimation, 240–243
freeze rates, 249
Fruchterman-Reingold algorithm, 261
functions, 33
    knn(), 81
    likelihood, 122
    loss, 68
    penalty, 160
fundamental problem of causal inference, 286

## G

garbage in, garbage out scenarios, 165
Gaussian distribution, 30
Gaussian mixture model, 56
Geller, Nancy, 8
Gelman, Andrew, 26
general functional form, 31
generating data, 306
generative processes, 52
geo-based location data, 23
geographic information systems (GIS), 220
geometric mean, 211
GetGlue, 135–137, 214
GetGlue exercise, 162–164
Giraph, 333
GitHub, 338
Gmail, 95
goodness, 301
Google, 2, 3, 4, 5, 21, 53, 126, 193
    Bell Labs and, 35

experimental infrastructures, 281
issues with, 176
machine learning and, 52
MapReduce and, 323
mixed-method approaches and, 194
privacy and, 196
sampling and, 21
skills for, 8
social layer at, 196
social research, approach to, 193–198
text-mining models and, 13
Google glasses, 6
Google+, 41, 193, 195, 253
graph statistics, 266
graph theory, 255
grouping data, 82
groups, 256
Guyon, Isabelle, 180, 181

## H

Hadoop, 21, 335–337
    analytical applications, 338
    Cloudera and, 337
    core components, 336
    MapReduce and, 335
Hammerbacher, Jeff, 8, 337
Hamming distance, 76
Hansen, Mark, 217–235
Harris, Harlan, 13
Harvard Business Review
    data science in, 9
HDFS, 336
heaps, 328
Hessian matrix, 123, 124
hierarchical modeling, 82
high bias, 192
high class imbalance, 242
Hippocratic Oath of Modeling, 356
Hofman, Jake, 342
Howard, Jeremy, 175
Hubway Data Visualization challenge, 251
Huffaker, David, 193
human powered airplane thought experiment, 334

likelihood function, 122
linear algebra, 207
linear regression algorithms, 30, 42,
    55–71
    evaluation metrics for, 66–70
    models, fitting, 61
    multiple, 68
    noise and, 64–66
    spam filters and, 95
    using, 55
linear regression model, 64
linear relationships, 55, 56
LinkedIn, 8, 253
    data science in, 9
Lives on a Screen (Thorp/Hansen),
    230
log returns vs. percent, 149
logistic regression, 54, 113–128
    algorithms, understanding, 117
    at runtime, 116
    classifiers, 115–118
    evaluating, 125–128
    implementing, 124
    in financial modeling, 158
    interpretability of, 117
    mathematical basis of, 120–122
    Media 6 Degrees (M6D) case
        study, 118–128
    Media 6 Degrees (M6D) exercise,
        128
    Newton's method of, 124
    output of, 119
    scalability, 117
    stochastic gradient descent meth-
        od, 124
    thought experiments for, 113–128
logistic regression model, 121
Lorenzian water wheel, 44
loss function, 68
lynx, 107
lynx --dump, 107

## M

machine learning, 52, 236
    challenges in, 242
    data visualization and, 236

    detecting suspicious activity, with,
        237–240
    Google and, 52
    models, productionizing, 246
machine learning algorithms, 52, 333
    domain expertise vs., 175
machine learning classifications, 204–
    206
    recommendation engines and,
        204–206
Madigan, David, 291, 346
Mahalanobis distance, 76
Mahout, 124
Manhattan distance, 76
MapReduce, 51, 326
    analytical applications, 338
    common usage of, 332
    Google and, 323
    Hadoop and, 335
    limitations of, 333
    using, 324, 330
    word frequency problems, 330
mathematical models, 27
matrix decomposition, 207
maximize information gain, 187
maximum likelihood estimation, 33,
    123
Mayer-Schoenberger, Viktor, 5
McKelvey, Jim, 236
MCMC methods, 268
mean absolute error, 127
mean squared error, 66, 67, 127, 309
meaning of features, 273
measurement errors, 203
Mechanical Turks, 168
    Amazon, 169
    crowdsourcing vs., 169
Mechanize, 107
Media 6 Degrees (M6D), 345
Media 6 Degrees (M6D) case study,
    118–128
    click models for, 118
Media 6 Degrees (M6D) exercise, 128
medical data thought experiment,
    292, 303
meta-definition thought experiment,
    13
Metamarket, 6

nodes, 136, 256
  pairs of, 258
noise, 65
noise in linear regression models, 64–66
normal distribution, 30
normalized Gini coefficient, 172
normalizing data, 148
null hypothesis, 67, 183

## O

Observational Medical Outcomes Partnership, 298–302
observational studies, 277, 283–289
  causal effect, 287
  causality, visualizing, 286–287
  improving on, 296–298
  in medical literature, 293
  Rubin Causal Model, 285
  Simpson's Paradox, 283–285
observations, 20
observed errors, 66
observed features, 81
OK Cupid, 276–279
orthogonal, 207, 208
overfitting models, 34, 203, 205
oversampling, 311
O'Neil, Cathy, 144

## P

p-values, 67, 183
Pandora, 137
parallelizing, 335
parameter estimation, 52
parameters, interpreting, 53
parsing tools, 107
  Beautiful Soup, 107
  lynx, 107
  lynx --dump, 107
  Mechanize, 107
  PostScript, 107
Patil, DJ, 8
patterns, 36
penalty function, 160
percent returns, 149
  log vs., 149
  scaled, 149

performance estimation, 240–243
Perlich, Claudia, 305–307, 345
Perlson, Doug, 46
personal data collection thought experiment, 219
Pinterest, 47
polishing, 341
populations, 19
  distinctions between samples and, 19
  super-, 22
position, 126
position normalize, 126
PostScript, 107
precision, 79, 127, 241
predicted preference, 210
predicting attributes, 204
predictions, 54, 57, 177, 273, 357
predictive models, 194
predictive power, 53
  interpretability vs., 192
predictors, 176
  adding, 68
Pregel, 51, 333
prescriptionists, 13
Principal Component Analysis (PCA), 206, 209–211
priors, 212
  higher derivatives and, 162
  in financial modeling, 159
privacy thought experiment, 197
privacy, data science and, 196
probabilistic models, 94
probability, 30, 119, 317
  binary outcomes vs., 317–320
probability distributions, 30–33
problems with channels, 328
process thinking, 339–341
processes
  data-generating, 19
  generative, 52
  real-world, 19
Processing programming language, 221
products, 5
proximity clustering, 261
prtobuf, 338

small-world networks, 268
social annotation, 196
social network analysis, 255, 260
social networks, 254–269
  analysis of, 255
  as school of fish, 262
  case attribute data vs., 254
  centrality measures, 257–259
  eigenvalue centrality and, 264
  Morningside Analytics, 260–263
  representations of, 264
  terminology of, 256
social research, 165–198
  extracting meaning from, 165–198
  feature selection, 176–193
sociometry, 255
spam classifiers, 119
spam filters, 93–112
  combining words in, 101
  for individual words, 99–101
  k-Nearest Neighbors (k-NN) and, 96
  Laplace smoothing, 103
  learning by example thought experiment, 93–97
  linear regression algorithms and, 95
  Naive Bayes, 98–102
span, 207
sparse vectors, 202
sparseness, 203
specific conditional entropy, 187
specificity, 78
Square, 236
  challenges of, 237
  data visualization at, 248–249
Square Wallet, 236
statistical inference, 18, 19, 66
statistical models, 26, 30, 52
statistics, 7, 8, 17–34
  epidemiology and, 293
  graph, 266
  journals, 293
  modeling, 26–34
  populations, 19
  samples, 19
  statistical inference, 18

stepwise regression, 181
  combined approach to, 182
  methods of, 181
stochastic gradient descent, 52, 124
stratifications, 289, 294
stratifying data, 82
subgroups, 256
subnetworks, 256
super-populations, 22
supervised learning, 81
  k-Nearest Neighbor algorithms, 71–81
  linear regression algorithms, 55–71
supervised learning recipe, 240
Suriowiecki, James, 169
Survival Analysis, 177

T

tacit knowledge, 359
Tarde, Gabriel, 218, 345
  Idea of Quantification, 218
Taylor Series, 183
teaching data science thought experiment, 350
test sets, 77
tests, 245
text data, 23
text-mining models, 13
TF-IDF vectors, 342
thought experiments
  access to medical records, 325
  automated statistician, 92
  chaos simulation, 44
  data science as a science, 114–128
  Egyptian politics, 259
  filter bubble, 213
  human powered airplane, 334
  image recognition, 108
  large-scale network analysis, 195
  learning by example, 93–97
  medical data, 292, 303
  meta-definitions, 13
  personal data collection, 219
  privacy, concerns/understanding of, 197

Robo-Graders, ethical implications of, 174
teaching data science, 350
timestamps in training data, 144
transaction data, 249
unified theory of data science, 114–128
thresholds, 300
Thrift, 338
time series modeling, 143
timestamped event data exercise, 162–164
timestamps
  absolute, 144
  financial modeling and, 137
  issues with, 142
timestamps in training data thought experiment, 144
tolerance, 329
traditional data, 23
traditional data fields, 220
training sets, 77
transaction data thought experiments, 249
transformations, 69
transforming data, 148
transforming features, 309
transitivity, 266
translations, 270
transpose, 207
treated, 275
trees, 320
trends, 60, 64
triadic closures, 256
triads, 256
true negative rate, 79
true positive rate, 79
true regression line, 65
Tukey, John, 34
Twitter, 21, 253, 259
  Lives on a Screen (Thorp/Hansen), 230
types of data, 23
  geo-based location, 23
  images, 23
  network, 23
  real-time streaming, 23
  records, 23

sensor, 23
text, 23
traditional, 23

## U

Ullman, Ellen, 40
unbiased estimators, 66
undirected networks, 264
unexplained variance, 67
unit level causal effect, 285
unit subsets, 20
unsupervised learning, 85
untreated, 275
usagists, 13
user retention example, 177–181
  interpretability vs.predictive power, 192
using MapReduce, 330

## V

variables
  correlated, 176
  random, 32
variance, 67
variation, 60, 64
vectors
  least important, 208
  TF-IDF, 342
Venn diagram of data science, 7
visualization, 8
visualization radiators, 248
volatility estimates, 153
volatility measurements, 153–155
Vowpal Wabbit, 124

## W

web, scraping data from, 106–108
Wikipedia, 167
Wills, Josh, 53, 334
Wong, Ian, 235–249
word frequency problems, 327–331
wrapper feature selection, 181–184
  algorithms, selecting, 181
  criterion for, 182
wrappers, 180

## X

xml descriptions, 235

## Y

Yahoo!
    Developer Network, 107

YQL Language, 107
Yau, Nathan, 7, 251

## Z

Zillow, 47

## About the Authors

**Cathy O'Neil** earned a PhD in math from Harvard, was a postdoc at the MIT math department, and a professor at Barnard College, where she published a number of research papers in arithmetic algebraic geometry. She then chucked it and switched over to the private sector. She worked as a quant for the hedge fund D.E. Shaw in the middle of the credit crisis, and then for RiskMetrics, a risk software company that assesses risk for the holdings of hedge funds and banks. She is currently a data scientist on the New York startup scene, writes a blog at *mathbabe.org*, and is involved with Occupy Wall Street.

**Rachel Schutt** is the Senior Vice President of Data Science at News Corp. She earned a PhD in Statistics from Columbia University, and was a statistician at Google Research for several years. She is an adjunct professor in Columbia's Department of Statistics and a founding member of the Education Committee for the Institute for Data Sciences and Engineering at Columbia. She holds several pending patents based on her work at Google, where she helped build user-facing products by prototyping algorithms and building models to understand user behavior. She has a master's degree in mathematics from NYU, and a master's degree in Engineering-Economic Systems and Operations Research from Stanford University. Her undergraduate degree is in Honors Mathematics from the University of Michigan.

## Colophon

The animal on the cover of *Doing Data Science* is a nine-banded armadillo (*Dasypus novemcinctus*), a mammal widespread throughout North, Central, and South America. From Latin, *novemcinctus* literally translates to "nine-banded" (after the telescoping rings of armor around the midsection), though the animal can actually have between 7 to 11 bands. The three-banded armadillo native to South America is the only armadillo that can roll into a ball for protection; other species have too many plates.

The armadillo's skin is perhaps its most notable feature. Brownish-gray and leathery, it is composed of scaly plates called *scutes* that cover everything but its underside. The animals also have powerful digging claws, and are known to create several burrows within their territory, which they mark with scent glands. Nine-banded armadillos typically weigh between 5.5 to 14 pounds, and are around the size of a large domestic cat. Its diet is largely made up of insects, though it will also eat fruit, small reptiles, and eggs.

Females almost always have a litter of four—quadruplets of the same gender, because the zygote splits into four embryos after implantation. Young armadillos have soft skin when they are born, but it hardens as they get older. They are able to walk within a few hours of birth.

Nine-banded armadillos are capable of jumping three to four feet in the air if startled. Though this reaction can scare off natural predators, it is usually fatal for the armadillo if an approaching car is what has frightened it, as it will collide with the underside of the vehicle. Another unfortunate connection between humans and nine-banded armadillos is that they are the only carriers of leprosy—it is not unheard of for humans to become infected when they eat or handle armadillos.

The cover image is from *Shaw's Zoology*, and was reinterpreted in color by Karen Montgomery. The cover fonts are URW Typewriter and Guardian Sans. The text font is Adobe Minion Pro; the heading font is Adobe Myriad Condensed; and the code font is Dalton Maag's Ubuntu Mono.